The Operators: The Simulation Hypothesis Physics, Philosophy, and Beyond

Robert S Kenyon

The Operators: The Simulation Hypothesis – Physics, Philosophy, and Beyond

Copyright © 2024 Robert S Kenyon

All rights reserved. No part of this publication may be reproduced, distributed, or transmitted in any form or by any means, including photocopying, recording, or other electronic or mechanical methods, without the prior written permission of the publisher, except in the case of brief quotations embodied in critical reviews and certain other non-commercial uses permitted by copyright law.

Independently published

ISBN: 9798343226799

This is a work of speculative non-fiction. While the concepts presented are speculative and philosophical in nature, references to real individuals, including theoretical physicists and their work, are based on factual information. Any interpretations or speculations are the author's own and are not intended to reflect the views of the individuals mentioned.

Printed in United Kingdom
First Edition

"Tron came first, maybe the Matrix is what they want you to believe."

This off-the-cuff remark I made on a Facebook thread was the spark that ignited my journey into the depths of theoretical physics, philosophy, and speculation. What began as a casual comment soon evolved into a profound exploration of reality, consciousness, and the universe itself. This book is the result of that journey—a quest to question the very fabric of existence and to consider the possibility that what we perceive as reality might just be the surface of something far more complex and elusive.

A Note to the Reader:

If you find it difficult to entertain the possibility that the universe could be a simulation, then this book may not resonate with you. The ideas presented here are highly speculative and unconventional, designed to push the boundaries of conventional thought and explore what might lie beneath the surface of our perceived reality.

This is not a book that seeks to convince you of the simulation hypothesis, nor does it claim to hold definitive answers to the universe's mysteries. Instead, it blends real, established physics theories with speculative thought, asking how our understanding of physics and quantum mechanics might change if the simulation theory were true. Some of the ideas are grounded in established science, while others challenge our current understanding in the spirit of exploration.

The goal of this journey is not to assert conclusions but to provoke curiosity and reflection. Could our interpretation of reality be shaped by human limitations and hubris, preventing us from seeing the universe's ultimate truth? Whether you agree or disagree, I hope that by the end, you'll have encountered perspectives that challenge and expand your thinking.

If you're open to questioning the nature of existence and blending established science with speculative inquiry, then

I welcome you on this journey. With this in mind, let us delve into the mysteries that have puzzled humanity for centuries, through a new and speculative lens.

Contents

The Operators 23

Introduction: The Operators 24

The 'How' vs. the 'Why': A Shift in Perspective 25

A New Lens on the Fermi Paradox 26

The Mechanics of the Simulation 27

The Quantum Universe: A Digital Framework 27

The Holographic Principle and the Nature of Reality 28

Consciousness and the Role of the Observer 28

Towards the Endgame: The Purpose of the Simulation 29

Chapter 1: The Cosmic Firewall – The Illusion of Isolation 30

The Fermi Paradox – A Misunderstanding of Cosmic Scale 31

Trapped in Time – The Limits of Cosmic Observation 32

The Cosmic Sandbox – Programmed Constraints on Civilizations. 34

Understanding the Scale – Time, Distance, and the Speed of Light. 35

The Speed of Light – The Cosmic Limitation on Communication 37

The Cosmic Firewall – The Laws of Physics as
Unbreakable Constraints. 39
The Cosmic Constraints: Speed of Light, Time, and
Entropy.. 40
The Laws of Physics as Programmed Firewalls. 42
The Purpose Behind the Firewall: Why Were These
Limits Imposed? 43
The Deception of Perception and the Boundaries of
Our Reality. 45
Black Holes and Wormholes – Cracks in the Firewall?
.... 48
Black Holes as Cosmic Recycling Mechanisms.......... 50
Wormholes: Backdoors in the Simulation? 51
Black Holes and Wormholes: Purposeful Design or
Glitches in the Matrix?... 53
Conclusion: Cracks in the Firewall 54
The Fermi Paradox Revisited – Programmed Silence... 56
A Silent Universe by Design. 57

Chapter 2: The Simulation Firewall 59

Cosmic Isolation and the Firewall. 60
Cosmic Isolation – A Programmed Design. 61
The Speed of Light – The Cosmic Firewall. 62
The Laws of Physics as Programmed Constraints..... 63

Dark Matter – The Hidden Framework. 65
The Purpose of the Firewall. 67
Worlds as Different Algorithms. 68
The Cosmic Purpose and the Unbreakable Firewall. 70
Conclusion: The Silent Cosmos Reconsidered. 71

Chapter 3: Wormholes and Black Holes: Backdoors and Information Systems 73

Cosmic Portals in the Simulation 74
Wormholes as Tools for the Operators. 74
Black Holes as Information Retrieval Systems. 76
The Black Hole Information Paradox and Hawking Radiation. 77
Quantum Information Theory and the Preservation of Data. 78
The Holographic Principle – Data Storage in the Universe. 79
Holographic Principle and the Black Hole Information Paradox. 80
The Holographic Principle in the Simulation Hypothesis 81
Black Holes as Data Nodes in the Simulation. 82
The Universe as a Hologram: A Layered Reality. 83

Philosophical Implications: Are We Living in a Hologram?.. 84

The Operators and the Mechanics of the Holographic Simulation. 85

Wormholes and Black Holes – Potential Escape Routes? 86

Quantum Mechanics, Information, and the Simulation. 88

Conclusion: Cosmic Backdoors and the Nature of the Simulation. 89

Chapter 4: Quantum Mechanics and the Observer Effect 91

Quantum Indeterminacy as a Simulation Rendering Process. 92

Rendering and the Role of the Observer. 93

Quantum Indeterminacy and the Boundaries of Reality. 94

Implications for the Nature of the Simulation.......... 95

Multiple Interpretations of Quantum Mechanics. ... 96

The Copenhagen Interpretation: Reality on Demand. 96

The Many-Worlds Interpretation: Infinite Simulations. 97

- The Pilot-Wave Theory: Hidden Variables and the Operators' Influence. 98
- Quantum Mechanics as a Multilayered Simulation: Unifying the Interpretations. 99
- Philosophical Implications: Is There a "True" Reality? 100
- Conclusion. 101

The Observer Effect and the Role of Consciousness. . 102
- The Observer Effect: A Brief Overview. 102
- Consciousness as an Interface: The Role of the Observer. 103
- The Participatory Universe: Wheeler's Delayed-Choice Experiment. 104
- Consciousness and the Rendering of Reality: A Deeper Connection. 105
- The Role of the Observer in the Simulation: A Hypothetical Scenario. 106
- The Ethical and Existential Implications. 107
- Conclusion. 108

Quantum Field Theory (QFT) and Its Relationship to the Simulation 108
- The Basics of Quantum Field Theory: A Brief Overview. 109

- Quantum Fields as the Building Blocks of the Simulation. 110
- The Role of the Vacuum: The Simulation's Background Code. 111
- Quantum Field Theory and the Operators' Control: Modifying the Code. 112
- Quantum Field Theory and the Nature of Reality: Is the Universe Just a Wave Function? 113
- QFT and the Multiverse: A Simulation of Infinite Possibilities. 114
- Conclusion. 115

Simplified Explanations and Analogies 116

- Quantum Indeterminacy and the Video Game Analogy. 116
- The Observer Effect and the Interactive Movie Analogy. 117
- Quantum Field Theory and the Wave on a Pond Analogy. 118
- The Quantum Vacuum and the Background Noise Analogy. 119
- Quantum Mechanics and the Coin Toss Analogy. .. 119
- Quantum Entanglement and the Paired Gloves Analogy. 120
- The Universe as a Giant Quantum Computer: The Library Analogy. 121

Summary and Reflective Thought................ 121

Chapter 5: String Theory, Quantum Engagement, and the Operators' Backdoor 123

Introduction. 124

A New Framework of Reality? 126

Quantum Entanglement as a Key Mechanism........ 126

Implications for Communication Across the Cosmos. 127

Beyond Space and Time: Quantum Systems as a Framework. 128

Quantum Manipulation and the Ultimate Escape? 129

String Theory and the Nature of Reality. 130

The Operators' Engagement with Reality............... 131

Quantum Engagement and the Purpose of the Simulation. 132

Quantum Entanglement and Cosmic Maintenance. 133

Quantum Entanglement and the Laws of Physics: A Balance of Control... 134

Quantum Entanglement and Communication with the Operators. 136

The Operators' Quantum Agenda........................... 137

The Quantum Universe as a Feedback System. 138

Quantum Entanglement and the Illusion of Free Will. 140

The Quantum Backdoor and Higher Dimensions. ... 142

Quantum Entanglement and the Question of Consciousness. 144

The Operators' Endgame: Quantum Revelation?... 145

Chapter 6: The Expansion of the Universe: Computational Analogies 147

Cosmic Expansion as a Sign of Increased Memory and Processing Power. 148

AI Scaling Laws and the Universe's Expansion. 148

Dark Energy: The Simulation's Optimization Algorithm. 149

Expansion and Increasing Complexity: The Role of Lifeforms. 150

The Role of Human Understanding and the Operators' Vision. 151

Entropy and Thermodynamics in a Simulated Environment. 152

Entropy as Information Dispersal: A Necessary Feature. 152

AI Regularization and the Second Law of Thermodynamics. 153

Entropy and the Operators' Experiment. 154

The Role of Lifeforms in Entropy Management...... 155

Entropy as a Measure of the Simulation's Progress.
.... 156

Computational Limits and Their Physical
Manifestations. 157

The Universe as a Finite Simulation: Boundaries of the
Sandbox. 157

The Planck Scale: The Universe's Pixel Resolution. 158

Physical Limits as Programming Constraints. 159

Pushing the Limits: What Happens When Civilizations
Test the Boundaries? 160

Physical Constants as the Operators' "Settings" 161

The Edge of the Sandbox: Are There Places We Can
Never Go? 162

Dark Energy as Software Updates Expanding the
Simulation. 163

Dark Energy: The Universe's Patch Notes............... 163

Expansion Mechanisms in AI Systems: Dynamic
Resource Allocation. 164

Dark Energy as a Safeguard Against Computational
Overload... 165

The Operators' Role in Implementing "Software
Updates" 165

The Implications of a Continuously Updated Universe. 166

The Endgame of Expansion: When Does It Stop?... 167

Philosophical Implications of an Expanding Simulation 168

Expansion as a Metaphor for Growth and Evolution. 168

Humanity's Role as Co-Creators in the Simulation. 169

The Role of Other Lifeforms: A Multitude of Perspectives. 170

The End of Expansion: Is There a Final Goal? 172

Our Place in the Expanding Simulation: Meaning and Purpose. 173

Chapter 7: Addressing the Fermi Paradox and Extraterrestrial Life 175

Revisiting the Fermi Paradox in the Context of the Simulation 176

Time Lags, Communication Barriers, and the Simulation's Constraints. 177

The Drake Equation and Probability of Life Within the Simulation. 178

Simulation Boundaries: Limiting Interaction Between Civilizations. 179

Cosmic "Sandbox Mode": Keeping Civilizations in Isolation.. 179

The Cosmic Firewall: Preventing Premature Contact. 180

Psychological and Cultural Barriers: Subtle Forms of Isolation.. 182

Simulation Boundaries as Protective Measures. 183

Testing the Boundaries: Are We Meant to Push Them? 184

The Role of Advanced Civilizations: Guardians, Guides, or Prisoners? 185

Expanding Consciousness: A New Form of Communication.. 186

The Ethical and Philosophical Implications of Using Altered States for Communication. 188

Expanding the Simulation: A Purposeful Expansion of Consciousness? 189

Conclusion: The Role of Expanded Consciousness in Overcoming Isolation. 189

The Operators' Purpose: Testing Isolation and Interaction.. 190

Isolation as a Catalyst for Innovation and Growth. 191

Interaction as a Catalyst for Understanding and Unity. 192

- The Endgame: What Happens When the Test is Complete? 193
- Transcending the Simulation: The Ultimate Goal? . 194
- Preparing for Contact: The Role of Culture and Imagination. 195
- The Operators' Legacy: What Will We Leave Behind? 196

Nick Bostrom's Simulation Argument: Implications and Expansions 199
- Expanding the Argument: The Role of the Operators. 200
- Implications for Reality and Existence. 201
- Ethical and Moral Implications. 202
- Conclusion: Bostrom's Argument as a Gateway to Deeper Questions 204

René Descartes and Scepticism of Reality: Parallels in the Simulation Hypothesis. 204
- Descartes' Method of Doubt and the Simulation Hypothesis 205
- The Role of the Observer: Consciousness in a Simulated World. 207
- Operators as the New "Evil Demon" 208
- Certainty and Doubt in the Simulation. 209

Conclusion: The Modern Relevance of Descartes' Scepticism. 210

Ethical Dimensions of Simulated Existence: Free Will, Purpose, and Moral Responsibility. 211

Free Will in a Simulated Universe: Are We Truly Free? 211

The Purpose of the Simulation: Experiment, Entertainment, or Enlightenment? 213

Moral Responsibility of the Operators: Gods or Guardians? 214

Human Ethics in a Simulated World: Does It Matter? 215

Conclusion: Navigating the Ethical Labyrinth of the Simulation Hypothesis. 217

The Role of Imagination in Speculating About Reality. 218

Imagination as a Tool for Philosophical Inquiry 218

Speculating Beyond the Observable Universe 219

Creative Storytelling as a Way to Explore the Hypothesis 220

Imagination and the Limits of Knowledge 223

Imagination as a Bridge Between Science and Philosophy. 224

Testing the Hypothesis: Searching for Anomalies in the Universe. 224

Computational Constraints and the "Resolution" of the Universe. 225

The Holographic Principle and Information Paradox. 226

Quantum Physics and Digital Reality. 226

Theoretical Limits and Practical Challenges. 227

Conclusion: The Power and Purpose of Imagination. 228

Chapter 9: Alternate Dimensions: Fiction, Physics, and the Simulation Hypothesis .. 230

Introduction: The Allure of Alternate Dimensions 231

Alternate Dimensions in Popular Culture: Expanding the Narrative Universe 232

Theoretical Physics and Alternate Dimensions: From String Theory to the Multiverse 234

String Theory and Higher Dimensions 237

The Birth of String Theory: Pioneers and Early Development 238

The Eleven Dimensions: The Second Superstring Revolution. 238

Brane Cosmology: Our Universe as a Membrane ... 239

Challenges and Controversies 240

Implications for the Simulation Hypothesis 241

The Multiverse Hypothesis. 242

Wormholes and Dimensional Gateways. 243

The Operators and the Multiverse: Creating and Managing Alternate Realities. 243

The Purpose of a Simulated Multiverse. 244

Cross-Dimensional Interactions and Bleed-Through. 245

Philosophical and Existential Implications: Living in a Multiverse 246

The Nature of Identity in a Multiverse 246

Moral Responsibility Across Dimensions. 247

Finding Meaning in a Simulated Multiverse. 247

Conclusion: Navigating the Multiverse of Simulations 248

Chapter 10: A Universe Designed by Code 250

Reality as a Program – Awaiting an Upgrade. 251

Programmable Parameters: A Future Upgrade? 252

Imaginable Modifications: Preparing for a New Reality. 252

Why Would the Operators Choose to Update? 253

An Ever-Evolving Simulation. 254

Can We Decipher the Code of the Simulation? 255

- Exploring Quantum Computing and Space-Time Manipulation.......... 257
- The Future of Physics: Beyond the Simulation. 259
- The Big Bang: Echoes of a Previous Update. 263
- A Universe Poised for Another Update?................. 265
- The Silent Universe Reconsidered. 266
- A Profound and Explosive Finish............................. 267

Final Thoughts: The Cosmic Experiment and the Human Experience 269

- Exploring the Boundaries of Reality: A Journey Beyond the Known 270
- Philosophical Reflections: On the Nature of Reality and Consciousness. 272
- The Human Perspective: Living in the Shadow of the Simulation. 273
- Final Reflection: What Lies Beyond the Firewall? .. 274
- Conclusion: A Call to Wonder 275

Dedicated to Maz Kenyon for being my late-night brainstorming buddy.

The Operators

Introduction: The Operators

Since the dawn of time, as humanity has gazed into the night sky, we've asked a question that transcends generations: Are we alone?

The Fermi Paradox poses an unsettling dilemma. With the vast number of stars and planets scattered across the cosmos, why haven't we detected any signs of intelligent life? Where are the other civilizations? Could we truly be the only beings capable of contemplating our existence?

The answer may not be as straightforward as we once thought.

What if our understanding of the universe is flawed? What if the universe isn't a chaotic expanse but a carefully crafted simulation—designed with purpose? And what if, beyond our comprehension, there are 'Operators'—entities guiding the flow of time, observing us, and controlling the reality we perceive? These Operators may not only watch but also shape the rise and fall of civilizations within the boundaries they've programmed.

Imagine the universe as a giant sandbox, a digital playground filled with galaxies, stars, black holes, and dark matter. Within this simulation, life emerges and evolves, raising questions that it was never intended to answer. Like characters in a complex program, we are confined to the rules imposed by the Operators: the speed of light, the passage of time, the limits of physics. These constraints are not random; they are intentional firewalls, designed to keep us from unveiling the full truth.

The 'How' vs. the 'Why': A Shift in Perspective

In the field of theoretical physics, much of the focus is on understanding how the universe works. How do black holes form? How does the speed of light limit communication across vast distances? How do quantum particles behave? These are all important questions that probe the mechanics of reality, but they stop short of asking the deeper, more profound question: why?

Why does the universe follow these rules? Why are phenomena like the speed of light or black holes so consistent and unyielding? The simulation hypothesis offers a different perspective—one that suggests these rules are not arbitrary, but deliberate. They exist for a reason, imposed by the Operators to serve a purpose we have yet to fully comprehend.

Scientists seek to explain how the speed of light acts as a barrier to interstellar communication, but perhaps we should be asking: why was this specific limit set in place? Could it be that the Operators designed this limitation to ensure that civilizations evolve in isolation, free from interference? Are the laws of physics merely part of a grander design, guiding us toward an end we cannot yet perceive?

This book shifts focus from the 'how' to the 'why,' proposing that the universe's governing laws—like the speed of light and quantum uncertainty—are not natural outcomes of randomness or chaotic forces. Instead, they are deliberate constructs, carefully crafted by the Operators to create the sandbox in which we exist. Scientific inquiry may help us uncover the how, but to truly understand the universe, we must explore the simulation hypothesis and grapple with the why.

A New Lens on the Fermi Paradox

We'll begin this journey by re-examining the Fermi Paradox, a mystery that has puzzled scientists for decades. When viewed through the lens of simulation theory, however, the paradox starts to unravel. Perhaps it isn't a paradox at all but rather a fundamental misunderstanding of the universe's true nature.

Maybe we aren't alone, but the constraints of time and space—imposed by the Operators—make it impossible for us to discover or communicate with other civilizations. They are out there, just like us, but they too are confined to their own cosmic silos, observing the universe through a lens that blinds them to the full reality.

The Mechanics of the Simulation

From there, we will delve deeper into the mechanics of this simulation: exploring black holes, dark matter, and the ultimate fate of the universe. Are these cosmic phenomena mere accidents of nature, or are they deliberately programmed features of the simulation, each serving a specific purpose? Could black holes be more than destructive forces—perhaps acting as the final collection points for all mass and information, feeding data back to the Operators as the simulation moves toward its conclusion?

The Quantum Universe: A Digital Framework

We'll also venture into the strange world of quantum mechanics, where particles can exist in multiple states simultaneously, and distant particles can communicate instantaneously through a phenomenon known as quantum entanglement. Could these quantum mysteries be clues to the underlying code of the simulation? Might the universe operate like a quantum computer, with the Operators using entanglement as a backdoor to interact with the simulation in real-time?

The Holographic Principle and the Nature of Reality

And what if the universe itself is a holographic projection, with all the information of our three-dimensional world encoded on a two-dimensional surface? This radical idea, supported by some physicists, suggests that what we perceive as reality is a mere projection of deeper, hidden layers of existence. The Operators, in this view, would be the architects of this projection, weaving the fabric of reality from a more fundamental source.

Consciousness and the Role of the Observer

We will also explore the role of consciousness in this grand design. Is our awareness merely a byproduct of the brain, or is it something more—a fundamental feature of the simulation itself? The observer effect in quantum mechanics, where the act of measurement seems to alter reality, hints that consciousness might play a more active role in the universe than we currently understand. Could the Operators have designed consciousness as a tool to shape the simulation, a way for the universe to observe and interact with itself?

Towards the Endgame: The Purpose of the Simulation

As we journey through these concepts, we'll return again and again to the central question: why? Why would the Operators create this universe? Are we part of a grand experiment, a cosmic test of intelligence, or a deeper exploration of consciousness itself? Or is the simulation an elaborate calculation, an attempt to reach an ultimate conclusion that only the Operators can comprehend?

By the end of this book, we may not have all the answers, but we will have explored the questions in a way that few have dared to before. We may find ourselves standing at the edge of a profound revelation—that the universe we inhabit is just one layer of a grander, more intricate design. And at the heart of it all, controlling the parameters of our existence, are the Operators—watching, waiting, and guiding us toward an end we can barely begin to imagine.

Chapter 1: The Cosmic Firewall – The Illusion of Isolation

The Fermi Paradox – A Misunderstanding of Cosmic Scale

The Fermi Paradox stands as one of the most perplexing questions in modern science. In a universe so vast—filled with billions of galaxies, each containing countless stars and potentially habitable planets—why have we not encountered intelligent extraterrestrial life? At first glance, this silence suggests that we might be alone in the cosmos. However, this assumption may reflect a profound misunderstanding of the nature of the universe itself.

What if this so-called paradox isn't a question of *where* or *when* alien civilizations exist, but rather a reflection of the constraints imposed upon us by the very framework of reality? Instead of asking why extraterrestrial civilizations haven't reached us, perhaps the more important question is whether the universe's fundamental laws are deliberately designed to keep us from ever discovering them.

The limitations of time, space, and the speed of light may not be arbitrary constants but carefully crafted barriers—a cosmic firewall, imposed not by natural forces but by the architects of a simulated universe. In this chapter, we'll explore the idea that our apparent isolation is an illusion, meticulously constructed to confine civilizations within their cosmic boundaries. The Fermi Paradox, then, may not be a reflection of alien absence but of programmed separation.

Trapped in Time – The Limits of Cosmic Observation

When we gaze into the vastness of space, we aren't looking at the universe as it exists in the present, but as it existed millions or billions of years ago. This is due to the finite speed of light—a fundamental boundary that dictates how fast information can travel across the cosmos. The very nature of this time delay creates a kind of temporal isolation between us and any other intelligent beings who may exist.

Imagine a civilization on a distant planet, equipped with the most advanced telescopes imaginable, looking in our direction. What would they see? They wouldn't observe the modern Earth with its cities, satellites, and technologies. Instead, they would see an ancient world—one potentially populated by dinosaurs, or even a barren planet before complex life evolved. In much the same way, when we peer outward into space, we are seeing ancient, primordial versions of distant worlds. The civilizations we seek might have long since evolved, or worse, they may have already vanished, leaving us with only a snapshot of their distant past.

This presents a fundamental challenge to the Fermi Paradox. The question assumes that if intelligent life exists, we should have found some evidence of it by now. But what if that evidence is already there, and we're just looking at the wrong point in time? What if those civilizations have evolved far beyond detection, or have ceased to exist altogether?

The cosmic time-lag ensures that we remain trapped in a kind of observational bubble, confined to seeing the universe as it was, not as it is. This creates the illusion of a lifeless universe, when in reality, we may simply be blind to the present state of distant civilizations. The Fermi Paradox isn't so much about the absence of extraterrestrial life; it's about the limitations of our current technology to find them in real-time.

However, is this limitation purely a consequence of cosmic scale, or could it be an intentional constraint? If the universe is a simulation, the speed of light might act as a boundary—a kind of artificial limit preventing us from real-time observation of distant civilizations. This cosmic sandbox keeps each civilization trapped in its own pocket of space-time, isolated from others. This deliberate programming could serve to maintain the integrity of the simulation, ensuring that no civilization evolves quickly enough to interfere with another's development.

The Cosmic Sandbox – Programmed Constraints on Civilizations.

In the world of computing, a sandbox is an isolated environment where actions are contained, ensuring that they don't interfere with the broader system. Could the universe itself be operating on similar principles? If we are indeed living in a simulation, the laws of physics may serve as the boundaries of our cosmic sandbox, ensuring that intelligent civilizations remain confined within their respective cosmic "bubbles."

Consider the speed of light. In our current understanding of physics, this represents the ultimate limit—a boundary that cannot be crossed. But what if this limit isn't just a physical constraint, but a deliberate programming choice? The sandbox of reality may have been designed to ensure that no civilization could develop the means to communicate or interact across vast cosmic distances, preserving the integrity of the simulation.

This isolation might serve several purposes. Perhaps it allows for a controlled experiment—an opportunity for the Operators to observe how intelligent life evolves independently of outside influence. Civilizations, in this model, are like test subjects in a grand cosmic experiment, each evolving in isolation within their own defined regions. The laws of physics act as walls, ensuring that no one can break free from their designated sandbox and interfere with the broader system.

It's also possible that this isolation is designed to limit technological advancement. While humanity dreams of faster-than-light travel, wormholes, and other means of bypassing the speed of light, these remain theoretical at best. The sandbox might be programmed so that no matter how advanced a civilization becomes, it will never achieve the breakthroughs needed to escape its cosmic bubble. This could be an intentional safeguard, preventing a civilization from growing too powerful or disrupting the balance of the simulation.

But what happens if a civilization manages to outsmart the system? Could the discovery of a way to bypass these constraints be part of the Operators' test? If so, the sandbox isn't just a limitation—it's a challenge, a barrier put in place to test whether intelligent life can transcend the limitations of its own reality. Civilizations that reach this point may be given the opportunity to break through the cosmic firewall and uncover the true nature of the simulation.

Understanding the Scale – Time, Distance, and the Speed of Light.

To fully appreciate the limits imposed by the speed of light, we need to consider the mind-boggling distances involved in space travel and communication. The numbers are so immense that they become almost incomprehensible, but breaking them down into more tangible examples can help us grasp their significance.

The speed of light—approximately 299,792 kilometres per second—is the fastest thing in the universe. Even at this incredible speed, light from the Sun takes 8 minutes and 20 seconds to reach Earth. When we look at the Sun, we're not seeing it as it is now but as it was over eight minutes ago. This might seem like a small delay, but it hints at the much larger time-lags that occur over greater distances.

To make this more relatable, consider how long it would take us to travel the same distance using current technology. A commercial jet flying at around 900 kilometres per hour would take nearly 19 years to travel from Earth to the Sun. A modern rocket, designed for maximum efficiency, would still take around 65 to 75 days to make the journey.

Now, let's scale this up. The nearest star system to our solar system, Proxima Centauri, is 4.24 light-years away. This means that even light—the fastest entity in the universe—takes more than four years to travel between our Sun and Proxima Centauri. Using today's spacecraft technology, it would take **tens of thousands of years** to cover that distance.

These examples highlight the incomprehensible vastness of space and underscore how limited we are by the laws of physics. Even the most advanced visions of space travel remain constrained by these boundaries, which, for all intents and purposes, seem unbreakable.

If intelligent life does exist near distant stars, by the time we receive any signals from them—if we ever do—those civilizations might have evolved far beyond what we can understand, or they may have ceased to exist entirely. The time it takes for light and radio waves to travel across these distances creates an almost unbridgeable gap, leaving us isolated in both time and space.

Could this limitation be intentional? If the universe is a simulation, the vast distances and time-lags could be part of the programming. The laws of physics might be designed to ensure that civilizations remain isolated from one another, limiting their ability to communicate, interact, or influence the broader system.

The Speed of Light – The Cosmic Limitation on Communication.

The time-lag imposed by the speed of light doesn't just delay our ability to observe the universe; it also fundamentally restricts how we can communicate with potential extraterrestrial civilizations. Let's imagine sending a radio signal to Proxima Centauri, the nearest star system. It would take over four years for the signal to arrive, assuming that someone there is able to receive it immediately. If they were to reply right away, we would have to wait another four years to receive their response, resulting in an eight-year round-trip conversation.

And this is for the closest star system. For more distant stars, galaxies, or planetary systems, the time-lag would stretch into hundreds, thousands, or even millions of years. Try to imagine having a conversation where each message takes millennia to reach its destination. By the time we hear back, entire civilizations might have risen and fallen, leaving behind only faint, ghostly remnants of their existence.

This inherent limitation leads us to an uncomfortable conclusion: even if advanced civilizations exist, it's highly probable that we will never be able to communicate with them in real-time. The speed of light creates a kind of temporal isolation that ensures that civilizations are effectively locked within their own regions of the universe, unable to engage with others in any meaningful way.

This brings us back to the Fermi Paradox. Could it be that the silence we perceive isn't due to a lack of life in the universe, but rather to this inherent cosmic isolation? The laws of physics themselves may be the barriers preventing us from bridging the gap between ourselves and other intelligent species. Even if we send signals into the void, the immense time-lags would make meaningful communication nearly impossible.

But here's a deeper question: Is this limitation simply a natural consequence of the universe's vast scale, or could it be by design? If the universe is a simulation, the speed of light may have been deliberately set as the maximum speed to ensure civilizations remain isolated, unable to influence one another. This programmed constraint would create a vast number of isolated pockets of life, each one evolving in ignorance of the others, ensuring that the simulation runs as intended.

The Cosmic Firewall – The Laws of Physics as Unbreakable Constraints.

In humanity's relentless quest to break boundaries, we have pushed the limits of knowledge and technology to extraordinary heights. We've built machines that can peer into the deepest corners of space, split the atom, and even decode the building blocks of life itself. Yet, despite all our advancements, certain barriers remain insurmountable. Chief among these is the speed of light—a cosmic speed limit that not only constrains our technological ambitions but defines the very boundaries of our universe.

The speed of light, approximately 299,792 kilometres per second, is not just an abstract figure in physics; it is a fundamental constant, a rule embedded in the very fabric of reality. As our understanding of physics deepens, it becomes clear that the speed of light represents more than just a physical limitation—it is a cosmic boundary, a designed firewall that confines us within the sandbox of the simulation. To grasp the full implications of this, we must explore how the laws of physics themselves act as programmed constraints, carefully crafted by the creators of the simulation to keep us within our cosmic limits.

The Cosmic Constraints: Speed of Light, Time, and Entropy.

To truly understand the nature of this cosmic firewall, we must first examine the key constraints that govern our universe: the speed of light, the passage of time, and the principle of entropy.

The speed of light is the most immediate and obvious of these constraints. According to Einstein's theory of relativity, as an object approaches the speed of light, its mass increases exponentially, requiring more and more energy to accelerate. As you near the speed of light, the energy required to keep accelerating becomes infinite, rendering faster-than-light travel impossible by any conventional means. No matter how advanced our technology becomes—no matter how efficient our engines or how vast our energy reserves—the laws of physics ensure that breaking this barrier remains out of reach.

However, the speed of light is not the only boundary keeping us confined. Time itself adds another layer to the cosmic firewall. The passage of time, which relentlessly moves forward, ensures that we are trapped in a linear progression, forever bound to experience the universe as a sequence of moments. Time moves in only one direction, erasing the past and obscuring the future, preventing us from interacting with other civilizations that may exist in different epochs of cosmic history. The idea of reaching back or forward in time remains a tantalizing dream—one that is perpetually out of reach.

Then there is entropy, the principle that governs the gradual decay of systems into disorder. In an isolated system, entropy always increases, which means that over time, the universe itself becomes less ordered, less structured, and increasingly chaotic. Entropy ensures that the universe moves toward greater disorder, making it nearly impossible to reverse or undo processes that might offer deeper insights into the nature of reality. In this sense, entropy acts as another layer of the firewall, preventing us from glimpsing the true nature of the universe by keeping the past irretrievable and the future unpredictable.

Taken together, these three forces—the speed of light, the relentless march of time, and the irreversible nature of entropy—create a system of nested constraints that isolate civilizations and limit our understanding of the cosmos. But could these limitations be more than mere physical constants? Could they be deliberate, programmed features of the universe, designed by the Operators to ensure that we never break free from our cosmic bubble?

The Laws of Physics as Programmed Firewalls.

In computing, a sandbox is an isolated environment in which actions are contained to prevent interference with the broader system. Could the universe itself be operating on similar principles? If we accept the simulation hypothesis, then the laws of physics—the speed of light, time, and entropy—may function as the parameters of our cosmic sandbox, carefully programmed by the Operators to keep intelligent civilizations confined to their respective regions of the universe.

Let's return to the speed of light as an example. In our understanding of physics, this speed represents the ultimate limit, a boundary we cannot surpass. But what if this boundary isn't just a physical constraint, but a deliberate programming choice? In this scenario, the speed of light acts as a firewall, preventing civilizations from communicating, interacting, or influencing one another across vast cosmic distances. It ensures that no matter how advanced a civilization becomes, it remains confined within its own cosmic region, unable to interfere with the broader system.

This isolation might serve several purposes. It could allow the Operators to conduct a controlled experiment, observing how intelligent life evolves independently, free from outside influence. Each civilization, in this model, becomes a test subject, evolving in isolation, governed by the rules set by the simulation. The laws of physics act as the walls of the sandbox, ensuring that no one can break free from their designated space and disrupt the grand experiment.

Moreover, these constraints might serve as a safeguard, limiting the technological advancement of civilizations. Even if we were to develop spacecraft with unlimited fuel or harness unimaginable amounts of energy, the laws of physics make it impossible to break the cosmic speed limit. The sandbox is programmed in such a way that no civilization will ever achieve the breakthroughs necessary to escape the confines of the universe. This limitation preserves the intended structure of the simulation, ensuring that the system runs smoothly, without interference from rogue civilizations that might destabilize the broader design.

The Purpose Behind the Firewall: Why Were These Limits Imposed?

At this point, the question shifts from *how* the laws of physics function to *why* they exist in the first place. If the universe is indeed a simulation, then we must ask: Why did the Operators impose these unbreakable constraints?

In a cosmic computer—the galactic-scale system that we inhabit—these laws serve as the foundational code that ensures the smooth operation of the simulation. The speed of light, the progression of time, and entropy are not arbitrary—they are the rules that keep the system balanced. Just as in any complex program, there are limits placed on what the software can do to prevent it from crashing or malfunctioning. In much the same way, the laws of physics are designed to maintain the stability of the universe, preventing chaos, overload, or system failure.

The speed of light, in this context, acts as the ultimate firewall, preventing civilizations from interacting in ways that could disrupt the simulation. It ensures that the sandbox remains isolated, with each civilization evolving independently within its own bubble of space-time. Without this limit, the simulation could become chaotic, with civilizations merging, sharing knowledge, and potentially overwhelming the system with their collective advancements.

Time and entropy serve similar functions. By trapping civilizations in a linear progression of time, the Operators ensure that no civilization can reach back into the past to alter the course of history or leap into the future to gain an unfair advantage. Entropy further reinforces this by ensuring that systems degrade over time, making it impossible to reverse the flow of information or recover lost data. These laws create a carefully balanced system, one that ensures the simulation runs smoothly, without the risk of collapse or interference from its inhabitants.

The Deception of Perception and the Boundaries of Our Reality.

Our perception of reality is inherently flawed. The solidity of objects, the sensation of touch, the distinction between the physical and the intangible—all these are illusions created by the interplay of electromagnetic forces. When we touch something, we never truly make contact. Instead, we are repelled by the electromagnetic fields of the atoms that make up both us and the objects around us. This creates a convincing illusion of solidity, but at the atomic level, most of what we perceive as matter is empty space.

If we could bypass these forces, we might be able to walk through walls, pass through what we consider to be solid objects. But this raises a deeper question: if our most fundamental experiences are deceptive, how can we trust our understanding of reality itself?

This deception of perception might not be a mere quirk of physics; it could be a feature of a far more sophisticated system designed to limit our comprehension of the universe—a cosmic firewall, programmed into the very fabric of our existence. Our current understanding of simulations and code is based on the rudimentary technology we possess—ones and zeroes, algorithms, and logic gates. These are the tools of a species only just beginning to grasp the nature of computation.

Imagine trying to explain modern digital technology to a society that has only ever known the abacus. The disparity is vast. Now imagine the gap between our technology and the technology of the operators, the creators of this hypothetical simulation. Their methods and systems would be as incomprehensible to us as quantum mechanics would be to an ant. The very concept of a 'simulation' could be as alien to them as their technology is to us. We might be living in a reality programmed not in binary but in a form of computational physics that melds spacetime, energy, and information into a seamless whole.

What we perceive as the universe—the laws of physics, the constants of nature, even the limits of our senses— could be constructs, constraints imposed by this advanced technology. They exist not to deceive, but to provide a consistent, navigable experience within the simulation. Our inability to truly touch, to see beyond certain spectra of light, or to move faster than the speed of light, might not be failings of our bodies or our understanding, but deliberate limitations set by the operators.

This idea is not unlike the boundaries within a video game. In any digital world, characters are confined by invisible walls, unable to stray outside the boundaries of the game's design. The only thing preventing a character from floating through every obstacle, through every wall or mountain, is the programmed code that defines the game's parameters. Remove these constraints, and the character would drift freely, unbound by the rules that once confined it. In this way, the boundaries are not flaws in the game but essential structures that create the illusion of a coherent world.

Similarly, our reality may be built on a code so advanced that we mistake it for the immutable laws of physics. The operators, like programmers of an unfathomably complex game, have set these constraints to maintain order, to keep us from wandering beyond the bounds of the simulation and into the unknown. These boundaries are not merely physical but conceptual, keeping our understanding tethered to what we perceive as 'real.' Our senses, our cognition, even the very fabric of our existence could be governed by this code, keeping us from perceiving what lies beyond the limits of the simulation.

This might explain why our current science and technology, impressive as they are, can only scratch the surface of understanding the universe. We are bound by the limitations of our tools and minds, much like characters confined within a game. We see the walls, we feel their solidity, but we can never move past them—not because it is truly impossible, but because the rules of the simulation say it is.

To explore this is to venture beyond the edges of our understanding, to question not just what is real, but what could be real. It challenges us to consider that the very framework of our universe—what we see, touch, and know—might be a beautifully constructed illusion, designed by beings with technology so advanced that it transcends our most ambitious speculations.

In this light, our perception is not merely a lie, but a necessary constraint, a filter that keeps us within the bounds of our limited understanding. The true nature of the universe, and the operators who created it, lies beyond these constraints, hidden in a realm we may never fully comprehend.

Black Holes and Wormholes – Cracks in the Firewall?

Black holes are among the most enigmatic and powerful entities in the universe. These cosmic phenomena represent points of such extreme gravitational pull that not even light—the fastest thing in the universe—can escape their grasp. When something crosses the **event horizon**—the boundary beyond which escape is impossible—it effectively vanishes from observable reality. In this way, black holes can be seen as the ultimate firewall, a barrier so absolute that anything drawn in seems to disappear forever, lost from view and beyond our reach.

For much of human history, this was our understanding of black holes: once something is swallowed by a black hole, it is gone forever, consumed by the immense gravitational forces at its core. This concept reinforced the notion of black holes as dead-ends in the cosmic landscape—vast, mysterious voids where matter, energy, and information are obliterated.

However, recent advances in theoretical physics suggest that black holes may not be the cosmic dead-ends we once thought. The **black hole information paradox**—a puzzle that has baffled physicists for decades—raises the possibility that information isn't destroyed within a black hole, but rather stored or encoded in some way. The concept of **Hawking radiation**, proposed by physicist Stephen Hawking, implies that black holes slowly emit particles over time, allowing them to gradually lose mass. This leads to the tantalizing idea that information might leak out from black holes, albeit slowly, rather than being lost forever.

These developments in our understanding of black holes challenge the notion of them as absolute firewalls. Instead, they may function as **cosmic data sinks**, collecting and processing the vast amounts of information generated by the universe. In the context of simulation theory, this raises an intriguing possibility: what if black holes are not just natural phenomena, but essential components of the simulation? What if they serve as conduits through which the Operators retrieve data from the cosmic sandbox, gathering the results of their grand experiment?

Black Holes as Cosmic Recycling Mechanisms.

In this view, black holes could be seen as a kind of **cosmic recycling mechanism**—a way for the simulation to process and store the immense amounts of information generated by the evolution of galaxies, stars, planets, and civilizations. Just as a computer periodically stores data or runs maintenance routines to keep its system stable, black holes might function as data collection points for the simulation. The matter and information consumed by black holes could be stored, processed, or even sent back to the Operators for analysis.

Imagine the simulation as a vast, interconnected network where every action, every event, and every decision made by intelligent civilizations generates data. Over time, this data must be gathered, processed, and stored somewhere within the system to ensure the simulation runs smoothly. Black holes could act as cosmic data repositories, collecting the information generated within the simulation and allowing the Operators to observe and study the progress of the various life forms and civilizations scattered across the universe.

Far from being the ultimate dead-ends we once thought, black holes could be integral to the structure of the simulation, providing a way for the system to maintain order and continuity. The information gathered and processed by black holes could offer the Operators insights into the evolution of the universe, the rise and fall of civilizations, and the overall functioning of the simulation. This process might be essential to the grand purpose of the simulation—an ongoing experiment to observe, analyse, and perhaps even optimize the progression of intelligent life.

Wormholes: Backdoors in the Simulation?

While black holes may act as cosmic data sinks, **wormholes** represent a more exotic and speculative possibility—one that hints at the existence of **backdoors** in the simulation. Though still theoretical, wormholes are believed to be passages through space-time that connect distant parts of the universe, potentially allowing information—or even entire civilizations—to travel faster than light and bypass the otherwise insurmountable limitations of space and time.

In the context of a simulated universe, wormholes could be much more than just a theoretical curiosity. They could represent cracks in the cosmic firewall, hidden shortcuts intentionally placed within the fabric of the simulation.

These **cosmic backdoors** might allow advanced civilizations to slip through the normally impenetrable barriers imposed by the laws of physics, such as the speed of light and the vast distances that separate galaxies.

If wormholes exist, they could provide a means of **bypassing the speed of light entirely,** allowing for near-instantaneous travel between distant points in the universe. For civilizations trapped within their own isolated regions of space-time, the discovery of a wormhole could be the key to breaking free from their cosmic confinement. In this way, wormholes could offer an opportunity to transcend the limitations of the simulation, revealing hidden pathways through which intelligent life might explore the broader system.

Much like black holes, wormholes could serve a vital function within the simulation. Rather than being anomalies or glitches in the system, they could be intentional **features** designed by the Operators—challenges or opportunities for civilizations advanced enough to discover and exploit them. Just as a computer program might have hidden codes or easter eggs designed to reward those who explore deeply enough, wormholes could serve as rewards for civilizations that push the boundaries of their knowledge and technology. They might represent a **test** within the simulation: a way for the Operators to determine whether intelligent life can break through the constraints imposed upon it and uncover deeper truths about the universe.

Black Holes and Wormholes: Purposeful Design or Glitches in the Matrix?

One of the most fascinating aspects of black holes and wormholes is the way they challenge our understanding of the universe's fundamental laws. Both phenomena seem to defy the conventional rules of physics, suggesting that the cosmic firewall may not be as impenetrable as it appears. In a simulated universe, this raises a critical question: are black holes and wormholes **deliberate features** of the simulation, or are they **glitches**—unintended cracks in the system that the Operators failed to account for?

On the one hand, the existence of black holes and wormholes could be seen as intentional design choices, placed within the simulation to offer civilizations the possibility of breaking through the barriers of time, space, and light. They might serve as challenges for intelligent life, encouraging civilizations to push the limits of their understanding and technology in order to discover the hidden workings of the universe.

On the other hand, black holes and wormholes might be **unintended byproducts** of the simulation—anomalies that arise from the complexity of the system itself. In this scenario, they could represent cracks in the cosmic firewall that were not part of the Operators' original plan. Perhaps black holes and wormholes offer a glimpse into the deeper structure of the simulation, revealing the limits of the Operators' control over the system. Advanced civilizations that discover and exploit these phenomena might be able to **circumvent the limitations** imposed by the simulation, accessing knowledge and technology that were never intended to be within their reach.

Whether black holes and wormholes are purposeful designs or unintentional anomalies, their existence suggests that the cosmic firewall is not entirely unbreakable. For civilizations that are advanced enough to understand and harness these phenomena, the possibilities are tantalizing. Black holes could be gateways to the ultimate purpose of the simulation, while wormholes might provide the means to explore and interact with distant regions of the universe, bypassing the carefully constructed limitations of the simulation's architecture.

Conclusion: Cracks in the Firewall

Black holes and wormholes represent cracks in the cosmic firewall—phenomena that challenge the traditional understanding of the universe's limitations and offer glimpses into the deeper structure of reality. Whether they are deliberate features or unintended anomalies, these phenomena suggest that the laws of physics may not be as absolute as we once thought.

In the context of the simulation hypothesis, black holes could serve as **cosmic data sinks**, collecting and processing the information generated by civilizations and delivering it back to the Operators for analysis. Wormholes, by contrast, could represent **backdoors** in the simulation— hidden pathways that allow advanced civilizations to break free from the constraints of time, space, and the speed of light.

For intelligent life trapped within the confines of the cosmic sandbox, the discovery and exploitation of these phenomena might offer the key to escaping the isolation imposed by the simulation. Black holes and wormholes may hold the secrets to unlocking the true nature of the universe, revealing what lies beyond the carefully constructed walls of the cosmic firewall. In doing so, they may offer the ultimate test for civilizations: can they transcend the limitations of their reality and uncover the hidden purpose of the simulation?

The Fermi Paradox Revisited – Programmed Silence.

Having explored the concepts of cosmic time-lag, the speed of light, and the isolation imposed by physical laws, it's time to revisit the Fermi Paradox. In light of the constraints we've discussed, the paradox takes on a new meaning.

The silence we perceive may not be a mystery at all, but rather a programmed feature of the universe. If the speed of light and other laws of physics are designed constraints, then it stands to reason that the universe was intended to keep civilizations isolated from one another. The Fermi Paradox isn't a question of why we haven't encountered extraterrestrial life—it's a reflection of the universe's structure, which prevents such encounters from ever taking place.

This leads us to the idea that the Fermi Paradox is not a paradox at all. The silence we experience is the direct result of programmed isolation, designed to ensure that civilizations evolve independently, free from external interference. By enforcing the speed of light as the ultimate limit, the Operators guarantee that intelligent species remain confined to their own bubbles of space-time, observing the universe in isolation.

But there may be exceptions. As we've discussed, the existence of black holes and wormholes suggests that there are cracks in the cosmic firewall. Advanced civilizations, if they are able to discover and exploit these phenomena, may find a way to bypass the isolation and transcend the limitations of their reality. This could represent the ultimate test for intelligent life within the simulation: the ability to break through the barriers imposed upon them and uncover the truth about the nature of their existence.

A Silent Universe by Design.

As we conclude this chapter, it becomes clear that the silence of the universe may not be a mystery to be solved, but a design choice—an intentional feature of the simulation. The creators of the universe, whether they are advanced beings or simply the architects of reality, have set the rules in place to ensure that civilizations remain isolated. The speed of light, the vast distances between stars, and the irreversible flow of time all serve to keep us confined within our cosmic sandbox.

However, the existence of phenomena like black holes and wormholes offers a tantalizing glimpse at the possibility of escape. These cracks in the cosmic firewall may provide the key to breaking free from the constraints imposed upon us, allowing civilizations to transcend the boundaries of their simulated reality.

In the following chapters, we will delve deeper into the nature of these cosmic backdoors, exploring how they might function within the simulation and what they could reveal about the true purpose of the universe. We will also consider what happens when civilizations reach the point of breaking through the firewall—what awaits them on the other side? Is there a greater reality beyond the simulation, or is the silence we hear simply the echo of a much larger, more complex game being played?

The questions ahead are vast and complex, but one thing is certain: the universe may be silent, but it is far from empty. The silence is by design, and it's up to us to figure out whether we can break through the walls that have been placed around us.

End of Chapter One

Chapter 2: The Simulation Firewall

Cosmic Isolation and the Firewall.

In the vastness of the observable universe, one of the most unsettling questions remains: why haven't we encountered extraterrestrial civilizations? If life is common, why are we alone? This perceived silence, often framed as the Fermi Paradox, may not be due to the absence of intelligent life, but rather the result of a deliberate firewall imposed by the architects of the simulation—the Operators.

This firewall serves multiple purposes. It ensures that civilizations, such as our own, remain confined within certain boundaries. It limits our ability to detect or contact extraterrestrial life, controls our technological advancement, and prevents us from uncovering the true nature of our reality. More than just a physical or technological limitation, the firewall is a fundamental aspect of the universe's coded structure, ensuring that each world, each civilization, follows a particular set of programmed rules or algorithms.

In this chapter, we explore the idea that the laws of physics, which we take as fundamental, are in fact programmed constraints designed to maintain the stability of the simulation. These constraints are not obstacles to be hacked or bypassed by advanced civilizations but are integral to the functioning of the cosmic computer itself.

Cosmic Isolation – A Programmed Design.

The absence of extraterrestrial contact, despite the staggering number of potentially habitable planets, leads to a startling conclusion: we are deliberately isolated. This isolation is not an accidental byproduct of the universe's vast size; rather, it is the result of a designed firewall. The Operators, whoever they may be, have ensured that civilizations remain in their own cosmic bubbles, operating under distinct rules and constraints, much like separate algorithms within a simulation.

From our perspective, we see the vast distances between stars and galaxies, the limitations imposed by the speed of light, and the apparent absence of signals from other civilizations. These are not random outcomes of natural laws—they are carefully crafted features of a controlled environment. The universe we perceive is, in fact, a cosmic sandbox—a place where civilizations develop independently, each following its own unique algorithm, without external interference.

Each living world within the simulation could be likened to a different algorithm, programmed to run its own course within the cosmic computer. Just as different software programs execute distinct tasks on a computer, each civilization may be governed by its own set of physical rules, limitations, and possibilities. The laws of physics that we experience are our specific algorithm, guiding the evolution of our technology, biology, and knowledge. Other worlds may have entirely different physical laws tailored to their unique purpose within the simulation.

This isolation serves several purposes. First, it protects the integrity of each civilization's development, allowing each to evolve without contamination from external sources. Second, it ensures that no single civilization can gain enough knowledge to break through the simulation's barriers. Each world, each algorithm, operates in isolation, advancing according to its own rules. The Operators use these boundaries to prevent civilizations from interacting or influencing one another, thus maintaining the sandbox environment necessary for their experiments or observations.

The Speed of Light – The Cosmic Firewall.

At the heart of this containment lies one of the most familiar physical laws: the speed of light. In our reality, the speed of light is the fastest that information can travel, creating an absolute limit on communication and observation. From the Operators' perspective, this limit functions as a key element of the firewall, preventing civilizations from communicating in real-time or traveling across vast cosmic distances.

The speed of light is not merely a physical constant; it is part of the underlying code of the cosmic simulation, a safeguard that maintains the integrity of each algorithm—each civilization. Imagine if we could travel faster than light, or communicate instantaneously with distant civilizations. Such advances would destabilize the very fabric of the simulation. The Operators' purpose, whatever it may be, hinges on keeping civilizations contained, developing in isolation. The speed of light is the perfect mechanism for this—its immutability ensures that no civilization can break the cosmic firewall.

Furthermore, the cosmic time-lag ensures that even if we do detect distant civilizations, we are seeing them as they were billions of years ago. The firewall is not just about space—it also applies to time, reinforcing the sense of cosmic isolation. The universe we see is a fragmented snapshot of ancient history, creating the illusion of an empty, lifeless cosmos. This perception is part of the Operators' design, ensuring that each civilization remains confined within its algorithmic bubble, unable to reach beyond the constraints of its coded reality.

The Laws of Physics as Programmed Constraints.
Beyond the speed of light, the fundamental laws of physics serve as additional layers of the simulation's firewall. These laws—the behaviour of matter and energy, the progression of time, and the forces that govern the universe—are programmed limitations designed to keep

civilizations contained within their respective simulations. From the Operators' perspective, they are akin to firewalls in a computer system, regulating access, controlling information, and ensuring the stability of the system.

Just as different algorithms within a computer are assigned specific parameters, each civilization within the cosmic simulation is governed by its own set of physical laws. These laws are not arbitrary; they are carefully coded rules that dictate the possibilities and limitations of that particular world. For instance, gravity, which holds galaxies together, and entropy, which ensures the one-way progression of time, can be viewed as elements of the simulation's code. These constraints are deliberately designed to create the experience of a coherent, stable universe, where everything behaves according to predictable laws.

In our case, the laws of physics function like the code that runs our algorithm, determining the shape and structure of our universe. These constraints prevent us from accessing certain levels of reality or discovering the deeper layers of the simulation. Just as firewalls in a computer prevent unauthorized access, the laws of physics ensure that we remain confined to our algorithmic bubble, unaware of the greater structure that lies beyond.

While we perceive these laws as inherent and immutable, they are, in fact, deliberate. Without these rules, the simulation would descend into chaos, with civilizations evolving unpredictably or potentially gaining access to the Operators' realm. The firewall of physics is essential to maintaining the illusion of a stable reality and ensuring that civilizations remain unaware of the true nature of the simulation.

Dark Matter – The Hidden Framework.

One of the greatest mysteries of modern physics is dark matter—a substance that we cannot directly observe, yet which appears to exert a massive gravitational influence, holding galaxies together. In our current understanding, dark matter interacts with visible matter through gravity, but it remains elusive, undetectable by any other means. But what if dark matter is more than just an unseen force? What if it is an essential part of the simulation's structural code—the hidden framework that maintains the stability of the universe?

In this hypothesis, dark matter functions as the unseen architecture of the simulation, ensuring the stability and coherence of everything we observe. Just as the casing of a computer holds its internal components together, dark matter operates as the "scaffolding" of the cosmic simulation, keeping the universe intact without interacting directly with the observable processes. It is the hardware that underpins the algorithms running throughout the simulation, directing the flow of information and ensuring that all systems remain stable.

This idea draws on theories such as John Wheeler's "It from Bit" philosophy, which posits that the universe is fundamentally informational—that all of reality is built from bits of data. If we accept this theory, then dark matter can be seen as part of the hardware layer that contains and directs the flow of this information. Much like how the casing of a memory card directs and contains the flow of electrons, dark matter directs the structural integrity of the universe, ensuring that the simulation runs smoothly.

Moreover, dark matter's invisibility could be intentional—its very purpose may be to serve as the hidden foundation of the simulation, inaccessible to civilizations like ours. Just as we cannot interact directly with the software code that runs a video game, dark matter could represent the hardware component of the cosmic computer that remains hidden behind layers of physics we are not equipped to breach.

Within the simulation, dark matter works in concert with the laws of physics—such as gravity and the speed of light—to maintain the stability of the cosmic sandbox. It acts as the silent, invisible structure that allows galaxies to hold together, planets to form, and civilizations to thrive, all while remaining outside the reach of direct detection or manipulation. This ensures that the Operators can stabilize the simulation while preventing us from fully understanding the mechanics of reality.

The Purpose of the Firewall.

To understand why the simulation firewall exists, we must consider the need for containment, control, and stability. In much the same way that a computer system requires specific components to channel the flow of information—like the plastic casing of a memory card that directs the flow of electrons—the universe requires structured systems to prevent chaos and disorder.

In this view, dark matter functions like the casing of the universe, creating a stable framework that holds galaxies and cosmic structures together. Without this framework, the universe would devolve into a chaotic storm of energy and matter, lacking any direction or control. The Operators, recognizing this, designed the simulation with structured limitations, ensuring that each world, each algorithm, evolves within carefully defined parameters.

The laws of physics, particularly the speed of light and gravity, serve as essential components of the *firewall*. These constraints restrict the flow of information and energy within the universe, preventing any civilization from gaining too much knowledge or power. By controlling how fast information can travel, how time progresses, and how matter interacts, the Operators ensure that the simulation remains stable and coherent, running according to the designed code.

This containment is not merely a means of limiting technological advancement—it is also a way to maintain the integrity of each civilization's algorithm. Just as different programs on a computer are kept isolated from one another to prevent interference, the cosmic firewall ensures that civilizations remain isolated, free from external contamination. Each world operates according to its own set of rules, its own version of the simulation, with the firewall keeping everything in check.

But the question remains: why? What purpose does this isolation serve? It may be that the Operators are running a grand experiment, testing how civilizations evolve under different conditions and algorithms. In this scenario, the firewall prevents civilizations from sharing knowledge or influencing one another, preserving the integrity of the experiment. Alternatively, the firewall could be a safeguard, ensuring that no civilization gains enough power to disrupt the entire simulation or access the Operators' realm.

Worlds as Different Algorithms.

As we look deeper into the simulation hypothesis, we begin to see the possibility that different living worlds—planets like Earth—could be akin to different *algorithms* running simultaneously within the simulation. Just as software programs operate under their own rules and parameters, each living world in the universe might be governed by its own unique set of physical laws, tailored to its specific algorithmic purpose within the cosmic computer.

In our case, the speed of light, gravity, and the flow of time form the core components of our algorithm. These laws determine the evolution of life, the development of technology, and the nature of our reality. However, on another world, the laws of physics might function differently. A world designed to test different parameters—perhaps a civilization that thrives on a fundamentally different understanding of matter and energy—could operate with its own version of the simulation code. On that world, faster-than-light travel might be possible, or the concept of time could flow in a nonlinear fashion.

The Operators, in this view, would be overseeing multiple algorithms simultaneously, each designed to explore different facets of life, intelligence, and progress. Some worlds might evolve along lines similar to our own, while others could follow radically different paths. The cosmic firewall, in this context, ensures that these algorithms remain separate, running in parallel without interfering with one another.

For us, the laws of physics are not just constraints to be escaped or hacked. They are fundamental components of our simulation code, designed to guide the progression of our civilization within its predefined boundaries. To break free of these constraints would not simply mean hacking the system—it would mean stepping outside the parameters of our algorithm altogether. The Operators have crafted a system where each world follows its own path, its own rules, ensuring that no civilization can easily transcend the boundaries of its reality.

The Cosmic Purpose and the Unbreakable Firewall.

At the heart of this cosmic design lies the question: why? Why were these constraints imposed? Why are civilizations isolated, each following its own set of rules?

The answer may lie in the very nature of the **cosmic computer itself**. The simulation firewall is not merely a protective barrier—it is an integral part of the system that ensures the smooth functioning of the simulation. By containing civilizations within their own bubbles, the Operators prevent them from destabilizing the broader simulation. The laws of physics are not arbitrary—they are carefully chosen to maintain order and control over the flow of information and energy.

In this sense, the firewall is unbreakable, not because civilizations lack the technology to overcome it, but because it is woven into the very fabric of the simulation's code. To break the firewall would be to unravel the simulation itself, creating chaos within the cosmic computer. Advanced civilizations may one day push the boundaries of their knowledge, but the Operators' design ensures that the firewall remains intact, preventing any civilization from stepping beyond the limits of its algorithmic reality.

This raises profound questions about the ultimate purpose of the simulation. Are we meant to remain within these constraints forever, or is there a way to transcend them? Could the Operators have left hidden pathways—backdoors—that advanced civilizations might one day discover, allowing them to glimpse the deeper layers of the simulation? Or are the laws of physics, dark matter, and the speed of light truly insurmountable, keeping us trapped within the cosmic sandbox for eternity?

Conclusion: The Silent Cosmos Reconsidered.

The idea that the universe is silent and empty may seem comforting to some, but unsettling to others. Within the context of the simulation firewall hypothesis, this silence is not the result of a lifeless cosmos but the consequence of deliberate design. The Operators have built the universe in such a way that civilizations remain isolated, each confined within its own algorithmic reality, unaware of the true nature of the simulation.

The firewall—comprising the speed of light, the laws of physics, and dark matter—ensures that no civilization can break through and discover the simulation's boundaries. Yet, the very existence of the firewall raises profound philosophical questions: Are we destined to remain trapped within this system forever, or is there a way out? Could there be hidden paths, wormholes, or backdoors placed there intentionally by the Operators, or are they vulnerabilities that advanced civilizations might one day exploit?

In the next chapter, we will dive deeper into the nature of wormholes and black holes, exploring how these phenomena might serve as the Operators' methods for accessing the simulation's hidden layers—or as potential escape routes for those civilizations that manage to unlock their secrets.

End of chapter 2

Chapter 3: Wormholes and Black Holes: Backdoors and Information Systems

Cosmic Portals in the Simulation

Throughout the universe, wormholes and black holes have fascinated scientists and philosophers alike. These mysterious phenomena seem to defy the very laws of physics, challenging our understanding of space, time, and information. In the context of the simulation hypothesis, wormholes and black holes take on an even deeper meaning—they may serve as the **backdoors** through which the Operators access the simulation or as **information retrieval systems** where vast amounts of data from civilizations are stored, processed, or retrieved.

In this chapter, we explore the role of wormholes and black holes as potential escape routes from the simulation, as well as tools for the Operators to manipulate or observe the information flow within the universe. These cosmic phenomena may be far more than just curiosities of physics; they may be essential components of the simulation's design.

Wormholes as Tools for the Operators.

Wormholes, theoretical passages through space-time, are often depicted in science fiction as shortcuts between distant points in the universe. But in the context of the simulation hypothesis, wormholes could be **tools deliberately placed by the Operators**. Rather than simply serving as physical shortcuts, wormholes may function as access points through which the Operators interact with the simulation.

Consider the idea that each world or civilization within the simulation follows a distinct algorithm, governed by specific physical laws and constraints. Wormholes, in this sense, might be the exceptions to those rules—the **backdoors** intentionally coded into the system. These cosmic portals could allow the Operators to **observe**, **reset**, or even **alter** the course of a civilization's development without being detected.

Furthermore, the nature of wormholes might vary depending on the algorithm governing each world. On some planets, wormholes might manifest as rare and transient phenomena, while on others, they could be stable and manipulable. If wormholes are tools of the Operators, their existence across different civilizations may represent an integral part of the **experiment**—an opportunity to test how different societies interact with these anomalies.

For civilizations that advance far enough to discover and exploit wormholes, the consequences could be profound. They might be able to access previously unreachable parts of the universe, or even break through the cosmic firewall that keeps them isolated. However, wormholes could also serve as **traps** set by the Operators, designed to limit how far a civilization can go before encountering the next layer of the simulation's complexity.

Black Holes as Information Retrieval Systems.

While wormholes may serve as **tools of access** for the Operators, black holes might function as **information retrieval systems**, collecting and storing the data generated by civilizations over time. In standard physics, black holes represent points of such extreme gravitational force that not even light can escape. Once something crosses the event horizon, it is lost forever—or so we thought.

In recent years, the **Black Hole Information Paradox** and the theory of **Hawking Radiation** have raised questions about whether black holes truly obliterate information, or if they merely store it in ways we don't yet understand. If we apply this to the simulation hypothesis, black holes might act as **cosmic hard drives**, gathering the information generated by civilizations and storing it for the Operators to analyse.

Imagine that, as civilizations progress, they produce vast amounts of data—scientific discoveries, technological advances, cultural milestones, and even their mistakes. When a civilization reaches a certain point of complexity, or perhaps when it collapses, all this information could be **retrieved** by black holes. The Operators might use these data points to observe the trajectory of different algorithms, compare civilizations, or even reboot certain aspects of the simulation.

Black holes, in this context, become more than just destructive forces. They are essential components of the simulation's information system, designed to preserve the data created by each world. The **event horizon** could be seen as the boundary between the visible universe and the deeper layers of the simulation, where information is stored beyond our reach.

The Black Hole Information Paradox and Hawking Radiation.

The **Black Hole Information Paradox** poses a fundamental question: what happens to the information that falls into a black hole? If the universe follows the laws of quantum mechanics, information cannot be destroyed. Yet black holes seem to erase anything that crosses their event horizon. The paradox lies in reconciling these two ideas.

Stephen Hawking's theory of **Hawking Radiation** suggests that black holes slowly emit particles, causing them to lose mass over time. This raises the possibility that black holes could eventually release the information they have absorbed, albeit in a highly degraded form. But in the context of the simulation hypothesis, this paradox takes on new meaning.

What if black holes are designed to **retrieve** information from the simulation and deliver it to the Operators? The information isn't lost or destroyed; it's simply **transferred** to another layer of reality, beyond our perception. In this sense, black holes act as **cosmic data repositories**, preserving the knowledge and experiences of civilizations and making them available to the Operators.

The **Holographic Principle**, which suggests that all the information contained within a black hole might be encoded on its surface, further supports this idea. Rather than falling into a singularity and disappearing, the information might be stored at the edge of the event horizon, forming a kind of **cosmic archive**. This would allow the Operators to access and analyse the information without disrupting the internal logic of the simulation.

Quantum Information Theory and the Preservation of Data.

At the heart of the Black Hole Information Paradox is the question of **how information is preserved** in the universe. In quantum mechanics, information is never truly lost. This principle suggests that even within the framework of the simulation, the Operators have built mechanisms to ensure that data from civilizations is stored and accessible.

Quantum information theory, which deals with the ways information can be encoded and preserved in quantum systems, offers insight into how the simulation might be designed to handle vast amounts of data. Black holes could represent one way the system **compresses** and **stores** information from multiple civilizations, ensuring that nothing is lost even as societies collapse or vanish.

In this view, quantum information is the fundamental building block of the simulation. Civilizations generate quantum information as they evolve, and this information is collected, stored, and potentially used to **reboot** or **reset** sections of the simulation. The Operators could use this data to refine algorithms, test new parameters, or even recreate entire civilizations after they collapse.

The Holographic Principle – Data Storage in the Universe.

The **Holographic Principle** is one of the most intriguing concepts in modern theoretical physics, and it offers profound implications for the **simulation hypothesis**. First proposed by Gerard 't Hooft and later expanded by Leonard Susskind, the principle suggests that all the information contained within a volume of space can be encoded on a two-dimensional surface, much like how a hologram encodes a three-dimensional image on a flat surface. This revolutionary idea has not only transformed our understanding of black holes but also provides a potential clue to the deeper nature of the universe—

especially when viewed through the lens of a simulated reality.

Holographic Principle and the Black Hole Information Paradox.

The **holographic principle** emerged as a potential solution to the **black hole information paradox**, which posed the question: what happens to the information about objects that fall into a black hole? If the laws of quantum mechanics hold true, information cannot be destroyed. However, black holes, with their immense gravitational pull, seem to erase everything that crosses their event horizon. The holographic principle offers a solution to this puzzle by suggesting that the information about everything that falls into a black hole is not lost but encoded on its **event horizon**—the boundary beyond which nothing can escape.

In this view, the black hole's event horizon functions like a **holographic storage device**, where all the information about the objects that fall into it is compressed onto the surface. The event horizon, therefore, acts as a two-dimensional boundary containing the total volume of information from within the black hole. This realization fundamentally shifts how we perceive not only black holes but space itself.

The implications of this principle stretch far beyond black holes. If information can be stored on a two-dimensional surface that encodes the entirety of a three-dimensional object, then space and matter themselves may be nothing more than **holographic projections** of underlying information.

The Holographic Principle in the Simulation Hypothesis.

If our universe is indeed a **simulation**, then the **three-dimensional reality** we experience might be nothing more than a **projection of underlying data** stored on a two-dimensional boundary. This echoes how a computer simulation generates complex, immersive worlds from relatively simple data structures. Just as video games use algorithms to create detailed 3D environments from lines of code, the universe could be a vast simulation running on a lower-dimensional data structure.

In the context of the simulation hypothesis, the **holographic principle** provides a critical clue to how the universe might operate. The three-dimensional world we observe could be generated by a **simulation algorithm**, with all the necessary data compressed onto a two-dimensional surface—the cosmic boundary of our universe. In this scenario, the **Operators**—the entities controlling the simulation—might have direct access to this underlying data and could manipulate it much like a programmer adjusting the code of a computer program to alter its output.

This perspective raises fascinating philosophical questions about the **nature of reality**. If the universe is a holographic projection, then the reality we perceive could be a mere **surface layer** of a far more complex system. The true information governing the universe might lie hidden in the depths of this **two-dimensional framework**. In essence, what we experience as the cosmos could be a **rendered interface**, while the real mechanics occur at a deeper, more fundamental level that is invisible to us.

Black Holes as Data Nodes in the Simulation.

The idea that black holes serve as **data nodes** within the simulation is a natural extension of the holographic principle. In this view, the **event horizon** of a black hole becomes a **storage device** where the information generated by the simulation—such as the outcomes of civilizations, stars, and galaxies—can be stored and compressed. The black hole doesn't destroy information; rather, it acts as a checkpoint within the simulation, preserving data that may be analysed or retrieved by the Operators at a later time.

In this sense, black holes serve as **centralized repositories** for vast amounts of data generated by the universe. The compression of information onto the event horizon allows the Operators to maintain control over the simulation's immense complexity. By storing and managing information in this way, the simulation can remain stable without requiring infinite amounts of processing power. Just as computer systems archive or compress data to save resources, black holes might play a similar role within the cosmic computer, ensuring that the simulation's data is safely stored and retrievable when necessary.

The Operators could access this information, modifying or resetting parts of the simulation as needed. Perhaps entire civilizations that have collapsed or reached a critical point in their development are encoded on the surface of black holes, waiting to be **rebooted** or **recreated** based on the Operators' will.

The Universe as a Hologram: A Layered Reality.

Taking this concept further, the **holographic nature of reality** suggests that the universe itself could be akin to a **holographic projection**—a simulation generated from lower-dimensional data encoded at the edges of the cosmos. In this model, the three-dimensional universe is simply the surface level of reality that we interact with, much like a video game's rendered graphics. But the true information about the universe—the underlying "code"—

is stored on a two-dimensional boundary that we cannot directly perceive.

This perspective aligns perfectly with the **simulation hypothesis**, as it offers a mechanism through which the Operators might store, compress, and manipulate the data of the universe. The Operators may not need to interact with the simulation on the level of the visible universe. Instead, they might adjust the information encoded on the **two-dimensional boundary**—the simulation's source code. By tweaking this code, they could influence or control the three-dimensional reality that we experience.

This idea also reinforces the possibility that our universe's boundaries are akin to a **data centre** or **hard drive** storing the entire history of the simulation. Every event, every star, every civilization could be encoded within this surface, with the Operators having the ability to retrieve and manipulate that information at will.

Philosophical Implications: Are We Living in a Hologram?

The holographic principle raises profound questions about the nature of reality. If our universe is a **holographic projection**, then everything we perceive—space, time, matter—might be an illusion, a manifestation of underlying data. Our understanding of the physical world is based on the assumption that what we see and experience is real, but the holographic principle challenges this assumption by suggesting that the entire universe could be a **simulated construct**.

In this context, the Operators might be able to control every aspect of reality simply by manipulating the information stored on the two-dimensional surface. Just as a computer programmer can alter a video game's code to change how the game behaves, the Operators could adjust the **simulation code** to influence the laws of physics, the behaviour of matter, or even the fate of civilizations.

Moreover, this leads to the unsettling possibility that the universe is not as **continuous** or **solid** as it appears. Just as a hologram is a projection of light from a two-dimensional surface, our experience of reality might be a projection of data stored on the edges of the universe. The boundaries between what is "real" and what is "simulated" become blurred, and the very concept of **material reality** begins to dissolve.

The Operators and the Mechanics of the Holographic Simulation.

The holographic nature of reality provides a compelling explanation for how the **Operators** manage the simulation. If all the information about the universe is encoded on a two-dimensional boundary, the Operators would have the ability to **access**, **modify**, and **delete** information at will. They could control the simulation with the same ease that a programmer controls the code behind a digital environment.

From this vantage point, black holes and their event horizons might be the most crucial nodes in the simulation's information network. The Operators could use black holes to **store** or **retrieve** data about different regions of the universe, civilizations, or even individual entities. Just as a computer system organizes data into folders or databases, the cosmic simulation might organize its vast amounts of information using **black holes** as storage units.

In this framework, the **Holographic Principle** serves as the fundamental basis for how the simulation operates—compressing complex information into more manageable forms and allowing the Operators to efficiently manage and control the universe without being constrained by its vastness or complexity.

Wormholes and Black Holes – Potential Escape Routes?

While we've discussed black holes and wormholes as tools for the Operators and data storage systems, the most intriguing question remains: could these phenomena also serve as **escape routes** for advanced civilizations?

In the simulation hypothesis, the **cosmic firewall** keeps civilizations isolated within their own regions of space-time, limiting their ability to communicate or interact with others. However, if a civilization were to become advanced enough to understand the mechanics of wormholes and black holes, it might be able to use them to escape the confines of its own simulation algorithm. This raises the possibility that wormholes could be **hidden backdoors**, allowing civilizations to bypass the simulation's constraints and explore other parts of the universe—or even other algorithms entirely.

For an advanced civilization, learning to manipulate wormholes could provide the key to **breaking through the simulation firewall**. By traveling through a wormhole, they might be able to access regions of the universe that were previously unreachable, or even bypass the rules of their own algorithm. In this way, wormholes could be more than just scientific curiosities—they could represent a potential **pathway to freedom** from the simulation itself.

Similarly, black holes might also serve as potential escape routes. While they are often viewed as destructive forces, recent advances in theoretical physics suggest that black holes might also act as **gateways** to other dimensions or layers of reality. If a civilization could learn to safely traverse a black hole, they might be able to access a deeper layer of the simulation—perhaps even reaching the realm of the Operators themselves.

Of course, these pathways would not be without risk. Black holes and wormholes represent some of the most extreme environments in the universe, and any attempt to use them as escape routes could result in catastrophic failure. But for civilizations that are advanced enough to understand and manipulate these phenomena, the potential rewards could be immense.

Quantum Mechanics, Information, and the Simulation.

Quantum mechanics introduces profound ideas about how information is handled in the universe—ideas that are central to the simulation hypothesis. The **observer effect**, quantum entanglement, and wave-particle duality all hint at the underlying informational structure of reality.

One of the most intriguing aspects of quantum mechanics is the **observer effect**—the idea that simply observing a quantum system can alter its outcome. This has led some to speculate that reality itself may not be fully "rendered" until it is observed. In the context of the simulation hypothesis, this suggests that the universe could be a **dynamically rendered simulation**, where only the parts of the simulation that are being observed or interacted with are fully processed.

This rendering process could also apply to **black holes and wormholes**. Perhaps they only fully exist when they are being interacted with or observed by a conscious observer. In this way, the simulation doesn't waste computational resources by fully rendering the entire universe all the time—it only renders the parts of the simulation that are actively being used.

The intersection of **quantum information theory** and the simulation hypothesis opens up a fascinating question: could the universe be designed in such a way that information is preserved and transferred not just through classical means, but through quantum processes? If so, wormholes and black holes could play an essential role in the **quantum information network** that governs the simulation. Civilizations that master quantum mechanics might be able to manipulate the very fabric of reality, accessing hidden layers of the simulation or even interacting with the Operators themselves.

Conclusion: Cosmic Backdoors and the Nature of the Simulation.

Wormholes and black holes represent some of the most mysterious and powerful phenomena in the universe. In the context of the simulation hypothesis, they may serve as **backdoors** through which the Operators access the simulation, as well as potential **escape routes** for advanced civilizations that have the knowledge and technology to use them.

As tools for the Operators, wormholes and black holes could be used to **retrieve information**, reset sections of the simulation, or alter the course of a civilization's development. But for those civilizations that manage to manipulate these phenomena, wormholes and black holes might offer a way to **escape the confines of the cosmic firewall** and access deeper layers of reality.

The universe, as a complex simulation, might be full of hidden pathways and backdoors that offer glimpses into the true nature of our existence. Whether these backdoors are intentionally placed by the Operators or represent vulnerabilities in the system, they hold the key to understanding the **deeper layers** of the simulation and the ultimate purpose of the cosmic experiment.

In the next chapter, we will delve further into the **quantum mechanics** behind the simulation, exploring how the observer effect and quantum field theory might offer additional insights into the mechanics of the cosmic computer.

End of chapter 3.

Chapter 4: Quantum Mechanics and the Observer Effect

Quantum Indeterminacy as a Simulation Rendering Process.

In the world of classical physics, the universe is predictable and deterministic. Every action has an equal and opposite reaction, and the trajectory of any object can be calculated precisely if we know its initial conditions. However, quantum mechanics presents a very different picture— one that seems almost chaotic by comparison. At the heart of this quantum world is the concept of **indeterminacy**.

Quantum indeterminacy posits that particles do not have definite positions or velocities until they are measured. Instead, they exist in a state of superposition, where all possible outcomes are equally probable. This is not just a limitation of our measurement tools; it is a fundamental feature of reality itself. But what if this indeterminacy is not just a quirk of nature but a deliberate feature of the simulation?

In a simulated environment, computational resources are not infinite. The operators of the simulation would need to allocate resources efficiently, rendering only what is necessary. This is similar to how modern video games only render the environment in the player's immediate vicinity, conserving processing power by not rendering distant objects until they are needed. Quantum indeterminacy could serve a similar function in the simulation, with particles existing in a state of probabilistic uncertainty until an observation, or measurement, is made—at which point the simulation "renders" a specific outcome.

This idea aligns with the **Copenhagen interpretation** of quantum mechanics, which suggests that a quantum system remains in superposition until it is observed, at which point the wave function collapses into a definite state. In a simulation, this collapse could be seen as the rendering process, where the simulation engine resolves the probabilities into a concrete reality based on the observer's input.

Rendering and the Role of the Observer.

The observer, in this context, plays a crucial role in determining what the simulation renders. When we look at a particle or measure its properties, we force the simulation to make a choice, to collapse the wave function and produce a specific outcome. This observer-dependent reality suggests that the universe, at its most fundamental level, is interactive and responsive—an intricate dance between the observer and the observed, where reality is not a fixed backdrop but a dynamic, evolving process.

This concept challenges our traditional understanding of reality. In the simulation hypothesis, the universe behaves like a sophisticated program that only expends resources when necessary. Quantum indeterminacy, then, is not a flaw or a mystery but a feature designed to optimise the functioning of the simulation, ensuring that it operates efficiently without wasting resources on unnecessary details.

Quantum Indeterminacy and the Boundaries of Reality.

One of the most striking implications of this idea is that reality, as we perceive it, may not exist independently of observation. Just as a video game only renders the parts of the world that the player can see, the simulation might only render the parts of the universe that are being observed. This means that the moon, to use a famous example, might not exist in a definite state when no one is looking at it. It only becomes real when observed, collapsing from a state of quantum potential into the tangible object we know.

This view also aligns with the **holographic principle**, which suggests that the entire universe could be described by information stored on a two-dimensional surface, with the three-dimensional world we experience being a kind of projection. In this context, quantum indeterminacy could be the mechanism by which this projection occurs, with reality being rendered in real-time based on the information encoded at the fundamental level of the simulation.

Implications for the Nature of the Simulation.

If quantum indeterminacy serves as a rendering process, it implies that the simulation is highly dynamic and interactive, with reality being shaped by the actions and observations of its inhabitants. This also means that the universe is not a fixed, pre-determined entity but a fluid, evolving system where the past, present, and future are constantly being shaped by conscious interaction.

In this light, the observer effect becomes more than just a peculiar aspect of quantum mechanics—it becomes a central feature of the simulation, a way for the operators to engage with and respond to the evolving state of the universe. It also suggests that consciousness itself might play a crucial role in shaping reality, acting as a kind of interface between the simulation and the inhabitants within it.

Quantum indeterminacy, far from being an enigmatic oddity, could be the key to understanding how the simulation operates. By rendering only what is observed, the simulation conserves resources, maintains consistency, and responds to the conscious experiences of its inhabitants. In this sense, the universe is not just a passive backdrop but an active participant in the unfolding of reality, with the observer playing a pivotal role in determining what is rendered and what remains in the realm of quantum potential.

Multiple Interpretations of Quantum Mechanics.

Quantum mechanics, with its counterintuitive principles and mind-bending implications, has spawned a variety of interpretations, each attempting to explain the bizarre nature of reality at the quantum level. These interpretations don't just describe how the universe operates—they offer potential insights into the very structure of the simulation. In this section, we'll explore the most prominent interpretations and how they might relate to the operators and their design of the simulated universe.

The Copenhagen Interpretation: Reality on Demand.

The **Copenhagen interpretation**, one of the oldest and most widely taught, suggests that particles exist in a superposition of states until they are observed. This means that reality, at its most fundamental level, does not exist independently but is **created by the act of observation**. From the perspective of the simulation hypothesis, this interpretation aligns perfectly with the idea that the universe renders reality on demand.

In a simulated environment, resources are not infinite. The operators would need to allocate processing power efficiently, rendering only what is necessary at any given moment. The Copenhagen interpretation could be seen as the **simulation's rendering protocol**: when an observer interacts with the system, the simulation resolves the probabilities into a concrete outcome, much like a video game rendering graphics only in the player's field of view.

This raises profound questions about the nature of existence within the simulation. If the universe only comes into definite form when observed, what does this mean for the parts of the simulation that remain unobserved? Are they mere potentialities, waiting to be brought into existence by a conscious observer? And if so, does this suggest that consciousness is a crucial component of the simulation, a tool used by the operators to bring the universe into being?

The Many-Worlds Interpretation: Infinite Simulations.

In stark contrast to the Copenhagen interpretation is the **Many-Worlds interpretation**, which proposes that every possible outcome of a quantum event actually occurs, each in its own separate universe. Rather than collapsing into a single reality, the universe continually splits into multiple, parallel realities where every possibility is realised.

From the standpoint of the simulation hypothesis, the Many-Worlds interpretation suggests a far more complex system, where the operators are not managing a single simulation but a **multiverse of simulations**. Every decision, every quantum event, spawns a new branch of reality, each with its own timeline and outcomes. This could imply that the operators are not just simulating one universe but are experimenting with an infinite array of possibilities, exploring every conceivable outcome across countless realities.

This interpretation challenges the notion of a singular purpose or endgame for the simulation. If every possible outcome exists, then the simulation might not be about reaching a specific conclusion but about **exploring the full spectrum of possibilities**. The operators, in this view, are less like experimenters seeking a particular answer and more like artists, creating an endless tapestry of realities, each with its own unique story.

The Pilot-Wave Theory: Hidden Variables and the Operators' Influence.

Another interpretation, less popular but no less intriguing, is the **Pilot-Wave Theory** (also known as Bohmian mechanics). It suggests that particles have definite trajectories, guided by a hidden wave function that exists in a higher-dimensional space. Unlike the randomness of the Copenhagen interpretation or the branching realities of Many-Worlds, the Pilot-Wave Theory implies a more deterministic universe, where hidden variables dictate the behaviour of particles.

For the simulation hypothesis, this theory could represent the **hidden code** of the simulation. The operators might use these hidden variables to control the simulation at a fundamental level, influencing the paths of particles and, by extension, the course of events. This hidden layer could be the operators' interface, a way to tweak the simulation without directly intervening in the observable reality. It suggests that what we perceive as randomness or uncertainty is, in fact, the result of the operators' influence, guiding the simulation according to their design.

Quantum Mechanics as a Multilayered Simulation: Unifying the Interpretations.

Each interpretation of quantum mechanics offers a different perspective on the nature of reality, and within the simulation hypothesis, they might not be mutually exclusive. It's possible that the universe operates on multiple levels, with each interpretation describing a different aspect of the simulation's architecture.

- The **Copenhagen interpretation** could apply to the rendering process, where reality solidifies in response to observation.
- The **Many-Worlds interpretation** might describe the branching structure of the simulation, with each branch representing a different scenario being explored by the operators.
- The **Pilot-Wave Theory** could represent the underlying code, the hidden mechanisms that guide the simulation and ensure that it follows the intended path.

In this view, the operators have crafted a multilayered simulation, using different quantum principles to manage various aspects of reality. The interpretations are not competing explanations but complementary components of a grander design, a sophisticated system that allows for both determinism and uncertainty, free will and predestination, randomness and order.

Philosophical Implications: Is There a "True" Reality?

If the simulation operates on multiple levels, each described by a different interpretation of quantum mechanics, then what is the true nature of reality? Is there a "base" reality, or are all levels of the simulation equally real? The operators, existing outside the simulation, would know the answer—but for those of us within it, the question remains tantalisingly out of reach.

This multilayered approach suggests that reality is far more complex than we can comprehend, with each layer offering a different perspective on the same underlying truth. For those within the simulation, understanding the full picture might be impossible. We are like characters in a video game, aware only of the world rendered around us, unaware of the complex code and hardware that make our existence possible.

This raises profound philosophical questions: If the operators exist outside the simulation, are they the true reality, or are they also part of a larger system? Is there an ultimate observer, a final layer of reality, or is existence an endless hierarchy of simulations, each nested within another?

These questions, while unanswerable, serve to highlight the depth and complexity of the simulation hypothesis. They remind us that our understanding of the universe is limited by the very nature of the simulation, and that the true nature of reality might be forever beyond our grasp.

Conclusion.

The various interpretations of quantum mechanics offer a glimpse into the possible architectures of the simulation.

Whether reality is rendered on demand, branching into infinite possibilities, or guided by hidden variables, each interpretation provides a unique lens through which to view the operators' design. In the next section, we will explore the role of the observer in more detail, examining how consciousness itself might be a key component of the simulation, influencing the very fabric of reality.

The Observer Effect and the Role of Consciousness.

One of the most intriguing and controversial aspects of quantum mechanics is the **observer effect**—the idea that the act of observation can influence the state of a quantum system. This phenomenon suggests that consciousness itself may play a fundamental role in shaping reality, and it opens the door to profound questions about the nature of existence, the function of the observer, and the purpose of the simulation.

The Observer Effect: A Brief Overview.
In classical physics, the world exists independently of whether or not it is being observed. An apple falls from a tree regardless of whether anyone is watching, and the planets orbit the sun according to the immutable laws of

gravity. But in the quantum realm, things are not so straightforward.

The observer effect was first brought to light through experiments such as the **double-slit experiment**, where particles such as electrons display wave-like behaviour when not observed but behave like particles when measured. This implies that the act of observation collapses the wave function, forcing the system to choose a definite state. Before observation, the system exists in a state of superposition, with all possible outcomes coexisting in a probabilistic blur.

From the perspective of the simulation hypothesis, this effect can be seen as a **rendering mechanism**. Just as a video game only renders what the player can see to conserve processing power, the simulation only "renders" a specific state when observed. This not only conserves the computational resources of the simulation but also creates a reality that is interactive and responsive, a reality that changes based on the input of its inhabitants.

Consciousness as an Interface: The Role of the Observer.

If the observer effect is real, then consciousness itself might be more than just a by-product of the brain—it could be an **interface** between the simulated universe and the underlying reality governed by the operators. In this view, consciousness is not just experiencing reality but actively participating in its creation.

This idea resonates with the philosophical concept of **idealism**, which posits that reality is fundamentally mental or consciousness-based. Within the simulation hypothesis, this could mean that the operators have designed consciousness to serve as a bridge between the physical simulation and the deeper, non-physical realm from which they operate. Through consciousness, the inhabitants of the simulation not only observe but also shape their reality, participating in the unfolding of the simulation's narrative.

The Participatory Universe: Wheeler's Delayed-Choice Experiment.

Physicist John Archibald Wheeler proposed the idea of a **participatory universe**, where the observer plays a key role in shaping the cosmos. His famous **delayed-choice experiment** suggests that the way we choose to measure a particle can retroactively influence its behaviour, even after it has passed through a slit or made a decision in its path.

This experiment challenges the notion of a pre-determined reality and implies that the universe is more like a participatory event, where the future is not set in stone and can be influenced by the actions of conscious beings.

From the simulation perspective, this could indicate that the simulation is not static but **interactive and evolving**, with the operators using consciousness as a tool to engage with and respond to the actions of the simulated inhabitants.

Consciousness and the Rendering of Reality: A Deeper Connection.

If consciousness is indeed a key component of the simulation, this raises questions about its nature and origin. Is it a purely biological phenomenon, arising from the complex interactions of neurons in the brain, or is it something more fundamental—a direct connection to the underlying structure of the simulation itself?

Some theories, such as **Integrated Information Theory (IIT)** and **Orch-OR** (orchestrated objective reduction), suggest that consciousness may be deeply tied to the fabric of reality, arising from complex systems or quantum processes in the brain. In the context of the simulation hypothesis, consciousness could be seen as a **programmed feature**, designed by the operators to allow inhabitants to experience and interact with the simulation in a meaningful way.

This would imply that consciousness is not an accident but a deliberate creation, a tool used by the operators to facilitate the interaction between the simulated universe and the minds of its inhabitants. It could even be that the purpose of the simulation is to explore the nature of consciousness itself, to understand how it emerges, evolves, and interacts with reality.

The Role of the Observer in the Simulation: A Hypothetical Scenario.

Imagine a scenario where the simulation is designed to evolve based on the input of its inhabitants. Every observation, every conscious experience, feeds back into the system, influencing the parameters of the simulation. The operators, in this view, are less like gods dictating the course of events and more like gardeners, nurturing the growth of a complex, self-organizing system.

In this scenario, the observer effect becomes a tool for **evolution and adaptation**. The simulation does not have a fixed outcome but is instead a dynamic process, shaped by the experiences and choices of its inhabitants. Consciousness, in this context, is the key that unlocks the potential of the simulation, allowing it to explore a vast array of possibilities and outcomes.

This could explain why the universe appears fine-tuned for life and consciousness. It is not that the parameters of the universe were set arbitrarily, but that they have evolved in response to the input of conscious beings. The operators, through the observer effect, allow the simulation to grow and change, adapting to the needs and desires of its inhabitants.

The Ethical and Existential Implications.

The idea that consciousness is a fundamental component of the simulation has profound ethical and existential implications. If we are active participants in shaping reality, then our choices, actions, and observations carry significant weight. We are not passive observers but co-creators of the universe, with the power to influence the simulation at a fundamental level.

This raises questions about the nature of free will, the responsibility of the observer, and the purpose of the simulation itself. Are we, as conscious beings, meant to discover and understand the nature of the simulation? Or are we simply here to experience, to observe, and to play our part in the unfolding story of the universe?

The operators, by designing consciousness as an interactive component of the simulation, may be inviting us to explore these questions, to seek out the deeper truths hidden within the fabric of reality. In this sense, the simulation is not just a passive environment but a living, breathing system, shaped by the collective consciousness of its inhabitants.

Conclusion.

The observer effect and the role of consciousness suggest that we are more than just inhabitants of a simulated universe—we are participants in its creation. Through our observations and experiences, we shape the reality around us, interacting with the simulation in ways that go beyond mere observation. In the next section, we will delve into the intricacies of quantum field theory and its relationship to the simulation, exploring how the operators might use these fields as the building blocks of the simulated universe.

Quantum Field Theory (QFT) and Its Relationship to the Simulation

Quantum Field Theory (QFT) is one of the most successful frameworks in modern physics, unifying the principles of quantum mechanics and special relativity to describe how particles interact at the most fundamental level. In QFT, particles are not seen as independent entities but as excitations or disturbances in underlying fields that permeate all of space. This perspective offers a unique way of understanding reality and, within the context of the simulation hypothesis, provides a compelling model for how the operators might have constructed the universe.

The Basics of Quantum Field Theory: A Brief Overview.

In classical physics, particles such as electrons, protons, and photons are considered distinct objects that interact through forces like gravity and electromagnetism. However, QFT tells a different story. According to this theory, the universe is made up of various **quantum fields**—one for each type of particle. These fields exist everywhere, even in the emptiest parts of space, and what we perceive as particles are actually just localized excitations of these fields.

For instance, an electron is not a tiny, solid object moving through space but rather a ripple in the **electron field.** Similarly, a photon is a ripple in the **electromagnetic field.** The interactions we observe between particles are actually interactions between these fields, governed by the principles of quantum mechanics.

From the simulation perspective, these quantum fields could be seen as the **underlying code** or structure of the universe. Instead of pixels on a screen or bits in a computer, the simulation is constructed from fields that encode the properties and behaviours of all matter and energy. The operators, in this view, are the programmers who have designed these fields and set the rules by which they interact.

Quantum Fields as the Building Blocks of the Simulation.

If quantum fields are the building blocks of reality, then the universe is not a collection of independent objects but a complex, interconnected web of fields that give rise to everything we see. This interconnectedness is reminiscent of a computer program, where every function and variable is part of a larger, integrated system. In the simulation hypothesis, the quantum fields could be the **subroutines** of the simulation, each responsible for a different aspect of the universe's behaviour.

For example, the **Higgs field** gives particles their mass, much like a variable in a program assigns value to a function. The **gravitational field** governs the curvature of space-time, akin to a physics engine in a video game determining how objects move and interact. Each field has its own rules and parameters, but they all work together to create the illusion of a seamless, coherent reality.

This raises the question: If the universe is a simulation, why use quantum fields at all? One possibility is that fields offer a way to create a highly detailed and dynamic simulation without needing to simulate every individual particle. By using fields, the operators can encode vast amounts of information into a few elegant equations, allowing the simulation to run smoothly and efficiently. It's a bit like using mathematical functions to generate complex landscapes in a video game rather than drawing every tree and mountain by hand.

The Role of the Vacuum: The Simulation's Background Code.

In QFT, the concept of the vacuum is not empty space but a seething, energetic field filled with virtual particles popping in and out of existence. This quantum vacuum is the ground state of the fields, the baseline level of energy from which everything else arises. It is both nothing and everything, the source of all particles and forces.

From a simulation standpoint, the vacuum could represent the **background code** of the universe, the fundamental layer of the simulation upon which everything else is built. Just as the operating system of a computer runs in the background, managing resources and executing commands, the vacuum might be the **operating system** of the simulation, maintaining the integrity of the quantum fields and ensuring that the rules of the simulation are followed.

This interpretation aligns with the idea that the universe is a **holographic projection**, with the information about the entire universe encoded on a two-dimensional surface. The quantum vacuum could be the canvas upon which this information is projected, the invisible framework that holds the simulation together. The operators, in this view, are not just programmers but also **system administrators**, maintaining the background processes that allow the simulation to function.

Quantum Field Theory and the Operators' Control: Modifying the Code.

If quantum fields are the building blocks of the simulation, then it stands to reason that the operators have the ability to modify these fields, altering the properties of the universe at will. This would be akin to a programmer tweaking the parameters of a program to achieve a desired outcome. For example, changing the value of the Higgs field could alter the mass of particles, potentially changing the structure of matter and the nature of physical laws.

Such modifications might not be arbitrary but could serve specific purposes, such as preventing certain outcomes or guiding the simulation in a particular direction. If the simulation is an experiment in the development of intelligent life, the operators might adjust the fields to create conditions that encourage the emergence and evolution of complex beings. Alternatively, they could use these modifications to correct errors or anomalies in the simulation, ensuring that it continues to function as intended.

This ability to modify the fundamental code of the universe would give the operators immense power, allowing them to shape reality at its most basic level. It would also mean that the laws of physics are not fixed but can be changed if necessary, providing a dynamic and adaptable framework for the simulation.

Quantum Field Theory and the Nature of Reality: Is the Universe Just a Wave Function?

One of the most profound implications of QFT is that reality, at its most fundamental level, might be nothing more than a complex wave function, a mathematical object that describes the probabilities of all possible outcomes. In this view, the universe is not a collection of objects moving through space and time but a single, unified wave function that encompasses everything.

From the perspective of the simulation hypothesis, this wave function could be the **master equation** of the simulation, the underlying code that generates the entire universe. The operators, as the creators of the simulation, would have access to this wave function, allowing them to predict and control every aspect of reality. They could use the wave function to run simulations within the simulation, exploring different possibilities and outcomes before choosing the path they want the universe to take.

This interpretation suggests that the universe is not a fixed, objective reality but a fluid, evolving process, shaped by the interactions between quantum fields and the choices of conscious beings. The operators, by manipulating the wave function, can influence the course of events, guiding the simulation towards their desired goal.

QFT and the Multiverse: A Simulation of Infinite Possibilities.

Quantum Field Theory, when combined with the Many-Worlds interpretation, suggests the existence of a **multiverse**—a vast, interconnected web of parallel realities, each with its own version of the quantum fields. In the simulation hypothesis, this multiverse could represent a vast array of simulations, each exploring a different set of parameters and outcomes.

The operators, in this view, are not managing a single simulation but a multiverse of simulations, each running simultaneously and interacting in complex ways. They could be using this multiverse to explore different possibilities, to test different scenarios, or to gather data on how different configurations of the quantum fields affect the development of life and consciousness.

This multiverse could also serve as a **sandbox** for the operators, a place where they can experiment with different versions of the simulation without affecting the primary universe. It would be a way to explore the consequences of changing the fundamental parameters of reality, to see how different configurations of the quantum fields play out before making changes to the main simulation.

Conclusion.

Quantum Field Theory provides a powerful framework for understanding the structure and dynamics of the universe, offering a glimpse into the underlying code of the simulation. The fields that make up reality are not just mathematical abstractions but the fundamental building blocks of the simulation, created and controlled by the operators. By manipulating these fields, the operators can shape the course of events, guiding the simulation towards their desired outcome.

In the next section, we will explore these ideas through simplified explanations and analogies, making the complex concepts of quantum mechanics and quantum field theory accessible to all readers, regardless of their background in physics.

Simplified Explanations and Analogies

Quantum mechanics and quantum field theory are notoriously complex subjects that can be challenging to grasp, even for those with a background in physics. To make these ideas more accessible, we'll use simplified explanations and analogies that relate these concepts to everyday experiences and familiar technology. This approach will help bridge the gap between the technical aspects of the simulation hypothesis and the broader audience.

Quantum Indeterminacy and the Video Game Analogy.

Imagine you're playing a video game with an expansive, open world. As you navigate through this world, the game only renders the parts of the environment that you can see. The mountains, trees, and characters in the distance don't exist in any detailed form until you get close enough for the game to render them. This is how the game conserves processing power, rendering only what's necessary for your experience.

Now, think of the universe as a similar kind of simulation. Quantum indeterminacy, where particles exist in a state of superposition until observed, can be compared to the game only rendering what the player sees. The particles remain in a state of potentiality, not fully realized, until an observation is made—just like the game only rendering distant objects when you approach them.

In this analogy, the observer is like the player, and the universe is like the game. The simulation renders the universe in response to our observations, making quantum indeterminacy not just a weird quirk of nature but a logical feature of a highly efficient simulation.

The Observer Effect and the Interactive Movie Analogy.

Imagine watching an interactive movie where the storyline changes based on the decisions you make. At certain points, you're asked to choose between different actions for the characters, and your choices affect the outcome of the story. Before you make a choice, all possible storylines exist as potential outcomes. But once you decide, the movie plays out according to your selection, collapsing all the other possibilities.

This is similar to the observer effect in quantum mechanics. Before observation, a quantum system exists in all possible states, like the branching storylines of the movie. But once an observation is made, the system "chooses" a single outcome, much like the movie following the path dictated by your choice. In this analogy, consciousness is not just a passive observer but an active participant, shaping reality by interacting with the simulation.

Quantum Field Theory and the Wave on a Pond Analogy.

Imagine a still pond on a calm day. If you drop a stone into the water, ripples spread out across the surface. These ripples represent disturbances in the water field, propagating outward from the point of impact. The ripples are not separate objects but are inseparable from the water—they are simply the water moving in a specific way.

In Quantum Field Theory, particles like electrons and photons are analogous to these ripples. They are not independent objects floating through space but excitations in their respective fields. The electron is a ripple in the electron field, just as the ripple in the pond is a disturbance in the water. This analogy helps illustrate the idea that everything we perceive as particles is actually a manifestation of underlying fields that fill the entire universe.

The Quantum Vacuum and the Background Noise Analogy.

Think of the universe as a quiet room. If you listen closely, you might hear a faint hum or background noise, even in the silence. This noise represents the **quantum vacuum**—a state that is never truly empty but filled with a constant, fluctuating background of virtual particles popping in and out of existence.

Just as the background noise is always there, even when you're not paying attention to it, the quantum vacuum is the ever-present backdrop of the universe. It's the ground state of all the quantum fields, the fundamental layer from which all particles and forces arise. This analogy helps convey the idea that what we perceive as empty space is actually teeming with invisible activity, a seething sea of energy that underlies all of reality.

Quantum Mechanics and the Coin Toss Analogy.

Imagine flipping a coin high into the air. As it spins, it's both heads and tails, a superposition of possibilities. But the moment you catch the coin and look at it, the outcome becomes definite—it's either heads or tails, not both. This is a simple way to understand the principle of superposition in quantum mechanics, where particles exist in all possible states until observed.

In the simulation hypothesis, the coin is like a quantum system, and the observer catching it is like the measurement collapsing the wave function. Before observation, the simulation allows all possibilities to coexist, conserving resources by not committing to a single outcome. Observation forces the simulation to "choose" an outcome, just as catching the coin reveals whether it's heads or tails.

Quantum Entanglement and the Paired Gloves Analogy.

Imagine you have a pair of gloves in separate boxes. If you open one box and find a right-hand glove, you instantly know that the other box contains the left-hand glove, no matter how far apart the boxes are. This is a simplified analogy for **quantum entanglement**, where two particles become linked, and the state of one instantly determines the state of the other, regardless of distance.

In the simulation hypothesis, entangled particles could be seen as paired gloves in different parts of the universe. The operators, through this entanglement, can coordinate actions across vast distances, maintaining the coherence and integrity of the simulation. This analogy helps illustrate the non-local nature of quantum entanglement, where information appears to travel faster than light, instantaneously connecting distant parts of the simulation.

The Universe as a Giant Quantum Computer: The Library Analogy.

Imagine a vast library filled with countless books, each representing a different possible state of the universe. Every time you make an observation, it's like pulling a specific book off the shelf and reading a page. The act of reading the page doesn't just inform you about the story; it also determines how the story unfolds. The book you chose affects which books are available for your next choice.

This is akin to the universe functioning as a giant **quantum computer**, where each observation is a computational step that influences the next. The operators might be the librarians, managing the books and ensuring that the story progresses in a coherent manner. This analogy helps convey the idea of the universe as a vast repository of information, constantly being updated and re-written based on the choices of its inhabitants.

Summary and Reflective Thought.

These analogies and simplified explanations provide a framework for understanding complex quantum concepts in an accessible way. They illustrate how quantum mechanics, quantum field theory, and the observer effect might fit into the simulation hypothesis, offering a glimpse into how the operators could manage and control the simulation. By viewing the universe as a dynamic, interactive system, where consciousness plays a key role, we gain a deeper appreciation for the mystery and wonder of existence.

If the universe is indeed a simulation, with consciousness as a key component, what does this imply about our role and purpose within it? Are we here merely to observe and experience, or is there a deeper reason why the operators have made us active participants in shaping reality?

End of chapter 4

Chapter 5: String Theory, Quantum Engagement, and the Operators' Backdoor

Introduction.

In our exploration of the simulation hypothesis, we've already considered the notion that the universe operates within a set of designed constraints—what I've called a *cosmic firewall*. One of the most prominent of these is the speed of light, which dictates the upper limit of information and physical travel within our observable reality. To travel between distant stars or galaxies within the confines of this limitation would require unimaginable timeframes, effectively trapping civilizations in their own corners of the cosmos.

But what if there is a system that transcends this? Enter **string theory** and **quantum engagement**—two pillars of modern theoretical physics that may provide a glimpse into the operators' true means of interacting with the universe.

String theory suggests that at the most fundamental level, reality is not composed of particles, but rather tiny, vibrating strings. These strings operate not just in the three dimensions of space and one of time, but across **up to eleven dimensions**, a concept that begins to blur the boundaries between science and what may seem like science fiction. Within this multi-dimensional framework, the limitations of the speed of light may be bypassed altogether.

Then, we encounter the enigma of **quantum entanglement**, a phenomenon Albert Einstein famously referred to as "spooky action at a distance." When two particles become entangled, their states are linked in such a way that changes to one particle are instantly reflected in the other, regardless of the distance between them. In this context, the speed of light becomes almost irrelevant, akin to walking speed in a world where instant teleportation is possible.

Could it be that **quantum entanglement** represents the method by which the operators of the simulation engage with the universe? While the laws of physics impose limitations for those of us bound by space and time, the operators, or whatever entity controls the simulation, could have access to this deeper quantum layer of reality—a realm where the traditional rules do not apply.

In a sense, quantum entanglement offers a *backdoor* into the workings of the universe. Information and influence can pass through this backdoor instantaneously, bypassing the laws of classical physics. For the operators, this would provide a means of controlling or observing the simulation without the constraints that govern our reality.

A New Framework of Reality?

String theory, with its multi-dimensional nature, may describe the very architecture of the simulation. If the speed of light is the top speed permitted within our four-dimensional experience, higher dimensions might allow the operators to "travel" between points in space or time in ways that would seem miraculous to us. Quantum entanglement could serve as a tool within that framework, enabling instant communication or manipulation across vast distances.

In this way, the limitations imposed by the speed of light are not a flaw or an absolute restriction. Instead, they are part of the sandbox—the rules designed to keep us from accessing these higher-dimensional tools or engaging in quantum manipulation ourselves. The operators, however, could transcend these boundaries at will, using systems such as quantum entanglement as a way to interact with the simulation from the outside.

Quantum Entanglement as a Key Mechanism.

Entanglement, as we know it, is a phenomenon that is difficult to harness or control. But from the perspective of the operators, it could be the primary mechanism for maintaining and observing the complex systems within the simulation. Imagine the universe as a quantum computer, with entanglement being the data stream that allows the operators to monitor and guide the simulation without interfering directly in the everyday flow of events.

This opens up the possibility that **we are observing quantum entanglement because it's a reflection of how the simulation itself functions**. We see particles become entangled, communicating instantaneously across great distances, because at the deepest level of the simulation's code, that's how information travels for the operators.

In this sense, quantum entanglement is not merely a curiosity of physics—it is the key that opens the door to a deeper, operator-level engagement with the simulation.

Implications for Communication Across the Cosmos.

Quantum entanglement also offers a new perspective on the **Fermi Paradox**. If other advanced civilizations exist, they may not be using traditional methods of communication that are bound by the speed of light. Instead, they could be engaging in **quantum communication**, utilizing entanglement to instantaneously exchange information across galaxies. However, unless we discover how to tap into this system ourselves, such communication would remain invisible to us, adding further weight to the idea that the Fermi Paradox arises from the inherent constraints of our designed universe.

Beyond Space and Time: Quantum Systems as a Framework.

If we consider the operators as entities capable of accessing and manipulating the simulation, it becomes clear that their toolkit must operate on a fundamentally different level from the constraints that bind those within the simulation. **Quantum systems**, and particularly quantum entanglement, might offer us a clue to how the operators circumvent the rules of space-time as we understand them.

In quantum mechanics, the traditional barriers of locality and causality seem to break down. **Entangled particles** can interact across vast distances instantaneously, and **quantum superposition** allows particles to exist in multiple states at once until observed. This fluidity of existence could be the very nature of the operators' interaction with the simulation. From their vantage point, the strict rules that govern our reality may be nothing more than **parameters within a program.**

This view aligns with the concept of **string theory**, where the universe is made up of vibrational patterns of strings in higher dimensions. In these dimensions, the operators might have the ability to navigate both space and time in ways unimaginable to us. Just as a programmer can move effortlessly between different lines of code, the operators could **leap through dimensions**, engaging with the simulation wherever and whenever they choose.

Thus, **quantum systems**, as strange and mysterious as they seem to us, may represent the very infrastructure the operators use to govern or observe their creation. Quantum mechanics, in this view, isn't a set of anomalies—it's a glimpse into the hidden layer of the universe, the **framework through which the operators maintain the simulation.**

Quantum Manipulation and the Ultimate Escape?

If quantum entanglement represents a kind of backdoor into the workings of the simulation, might there also be a way for those within the simulation to exploit this system? Could we, in theory, find a way to overcome the sandbox limitations imposed by the speed of light and other physical laws by learning to harness quantum phenomena on a large scale?

At present, the technology to manipulate quantum systems for macroscopic purposes—such as communication or travel—remains beyond our reach. However, if such technology were developed, it could offer a way to **escape the cosmic firewall**. Advanced civilizations, in this scenario, may have already discovered this, utilizing **quantum manipulation** to travel vast distances instantaneously, bypassing the limitations of space-time.

For now, we are bound by our technology and understanding, but the potential exists that one day we might learn to exploit the loopholes in the simulation's design. In doing so, we could move beyond the constraints that define our current existence and possibly **access the level of control reserved for the operators.**

String Theory and the Nature of Reality.

String theory presents a universe that is vastly more complex than the one we observe, with **multiple dimensions** hidden from our perception. These dimensions, which might range up to eleven in total, offer tantalizing possibilities for the nature of reality itself.

Could these extra dimensions represent the **operating system of the simulation**, layers of existence where the operators function without the constraints of our observable reality? If the traditional four dimensions are simply the user interface, then the higher dimensions could be where the real action happens—the place where **quantum mechanics**, **entanglement**, and **information processing** take on a deeper meaning.

It's possible that within these hidden dimensions, **distance and time become irrelevant**, allowing information to travel instantly, unrestricted by the speed of light. The strings of string theory may act as **vibrational threads** that connect every point in the simulation, forming a vast, unseen network through which the operators can monitor, adjust, and manipulate the universe as needed.

The Operators' Engagement with Reality

If quantum entanglement allows for instantaneous information transfer, it's plausible that the operators use it to monitor the simulation without directly intervening. Much like a **central processing unit** in a computer, the operators may not need to physically "touch" any part of the simulation to keep it running. Instead, they can observe and influence outcomes through the entanglement of particles and systems across vast scales.

Quantum entanglement could be a means of **constant surveillance**, ensuring that the simulation operates within the desired parameters. Through this quantum layer, the operators would have real-time feedback on the state of the universe, without being limited by light-speed delays or spatial distances. It's as if the entire universe is **interconnected by a web of entangled particles**, acting as conduits for information flow.

This creates a reality where, from our perspective, the universe appears bound by time, space, and causality, while from the operators' viewpoint, it is a fluid, manipulable structure where information and matter are malleable. In this sense, quantum entanglement is more than just a peculiar phenomenon—it is the **primary tool of engagement** for the operators within the simulation.

Quantum Engagement and the Purpose of the Simulation.

As we explore this possibility, another question arises: what is the ultimate purpose of the simulation, and how do string theory and quantum systems relate to this goal? Could it be that the **quantum entanglement we observe is a by-product of the simulation's true function**—a function that may be far beyond our current comprehension?

If the simulation is designed to gather information, as hypothesized earlier, quantum engagement could be a way to efficiently process and transmit that information across vast distances. The **entanglement of particles** might be part of a sophisticated system of data management, ensuring that every event, every interaction, and every outcome is **instantly recorded and processed** within the simulation. In this view, quantum entanglement is the *nervous system* of the simulation, constantly relaying information back to the operators, allowing them to refine and perfect the model.

Perhaps the ultimate aim of the simulation, as with any experiment, is to gather data that will eventually lead to a specific outcome or discovery. The laws of physics, the constraints on faster-than-light travel, and the observable universe may all be **designed limitations**, keeping us focused on a specific path, while the operators work towards this final outcome through their own quantum manipulations.

Quantum Entanglement and Cosmic Maintenance.

If we consider quantum entanglement as the primary tool of the operators, we can begin to speculate about its role in the ongoing **maintenance and stability of the simulation**. Much like how a computer system requires constant monitoring, patches, and updates, the universe might require ongoing intervention to remain functional and to avoid 'bugs' that could disrupt the flow of events.

Quantum entanglement could serve as the **maintenance system** for the entire simulation. Through entangled particles, the operators may be able to detect irregularities, shifts in energy, or deviations from the intended parameters. These entangled systems might act like **sensors or monitors**, transmitting information back to the operators instantaneously, enabling them to correct errors in real-time.

From our vantage point, these corrections might manifest as what we perceive to be **quantum fluctuations** or other oddities that scientists currently find difficult to explain.

We may observe randomness or uncertainty in quantum mechanics because, at this deeper level of reality, the simulation is constantly being adjusted, fine-tuned, and optimised by the operators.

This leads to the idea that quantum randomness is not truly random—it could be **deliberate noise** inserted by the operators to conceal the underlying system from the inhabitants of the simulation. Much like the unpredictable behaviour of a computer program that has been designed to appear organic, the quantum world might be a veil, hiding the precision and intention behind its construction.

Quantum Entanglement and the Laws of Physics: A Balance of Control.

Quantum entanglement also suggests a **delicate balance** in how the operators manage the laws of physics within the simulation. The classical laws of physics—such as gravity, thermodynamics, and the speed of light—represent the rigid framework that keeps the simulation running smoothly, ensuring stability and predictability. These laws act as **fixed boundaries**, preventing chaotic or disruptive behaviours that might otherwise destabilise the universe.

But then we encounter quantum mechanics, a field where those classical laws seem to break down. At the quantum level, particles can be in two places at once, entangled particles communicate instantaneously, and uncertainty reigns. It's tempting to view this as a flaw or paradox within the system, but perhaps it's not. Perhaps the **operators designed it this way**, blending **classical control** with **quantum flexibility** to create a dynamic system that could evolve without becoming chaotic.

Quantum entanglement may be the **bridge** between these two realms—an invisible tether that allows the operators to intervene in the simulation without disturbing the classical laws. The entangled particles, like hidden threads, enable the operators to pull at the very fabric of the universe, subtly guiding it in the right direction. By doing so, they maintain a balance between the fixed laws of the simulation and the need for flexibility and growth.

Are We Observing the Operators' Tools?

When physicists experiment with quantum mechanics and entanglement, they might be unknowingly **interacting with the operators' tools**. Every time we observe a particle behaving in a strange or unexpected way, every time we witness entangled particles responding instantaneously to each other, we are possibly seeing **the underlying system at work**—a glimpse behind the curtain of the simulation.

In this way, quantum physics becomes not just a study of the smallest building blocks of the universe, but a window into the mechanisms that keep the simulation running. **What we call entanglement** could be the operators' method for manipulating reality, and by experimenting with it, we might be stumbling upon the architecture of the simulation itself.

If this is true, then we are closer to the operators than we may realise. Our advancements in quantum physics could eventually lead to a deeper understanding of how the simulation works, perhaps even offering a way to **communicate with or manipulate the simulation** on our own terms.

Quantum Entanglement and Communication with the Operators.

One of the most tantalising possibilities of quantum entanglement is that it may offer a method for us to **communicate with the operators**. If the operators are using entangled particles to monitor and engage with the simulation, then perhaps there is a way for us to tap into this network.

Consider this: just as we send signals through a computer network to interact with a system, the operators might be sending **quantum signals** through the universe, encoded within the entanglement of particles. If we could decipher these signals or learn to send our own quantum communications, it might be possible to make contact with the operators.

The challenge, of course, is that the **language of the operators** is encoded in the very fabric of quantum mechanics. Our current understanding of quantum entanglement is limited, but as we push the boundaries of quantum computing, quantum cryptography, and entanglement-based communications, we may one day find a way to tap into the **quantum channels** that the operators use to control the simulation. This would represent a profound breakthrough, allowing us to communicate beyond the confines of space-time and perhaps even **alter the parameters** of the simulation itself.

Could this be the future of human progress? A world where we are no longer bound by the speed of light, where we communicate and interact on a quantum level with the very entities that created the simulation?

The Operators' Quantum Agenda.

If quantum entanglement is the tool of the operators, what then is their ultimate purpose? As we look deeper into the workings of the universe, we begin to realise that the operators' agenda may not be something we can comprehend in full. It is clear, however, that their use of quantum systems is not random—it is precise, deliberate, and deeply ingrained in the functioning of the universe.

Perhaps the operators are **testing the simulation**, observing how complex systems evolve within the constraints they've established. Or perhaps they are **waiting for us** to reach a level of understanding that allows us to transcend the physical limits of our reality and join them in a new phase of existence, where the classical laws of physics no longer apply.

The presence of quantum entanglement might signal that the simulation is not static—it is **interactive**, designed to respond to the progress of its inhabitants. The operators may be using quantum systems to guide us toward a deeper understanding of the universe, nudging us towards breakthroughs that will eventually allow us to escape the sandbox entirely. In this sense, quantum entanglement becomes not just a backdoor for the operators, but a **path for us to follow**—a way to engage with the simulation at its deepest level.

The Quantum Universe as a Feedback System.

If quantum entanglement serves as the **invisible framework** connecting all things, then it stands to reason that the universe is not a static simulation but a **feedback loop**. Every event, every decision made by conscious beings, every quantum interaction could send ripples through this quantum network, transmitting data back to the operators.

In this light, the universe operates much like a **self-regulating system**—constantly refining, evolving, and responding to the inputs provided by those within it. This feedback system would allow the operators to track the progress of civilizations, observe how matter and energy behave, and fine-tune the parameters of the simulation in real-time.

Imagine, for instance, a simulation where every decision made by a sentient being sends **quantum signals** through entangled particles, instantly updating the state of the universe. This could provide the operators with a wealth of data on how different conditions (such as varying physical laws, the emergence of intelligent life, or environmental pressures) affect the development of the simulation. The operators could then use this data to **adjust the simulation's code**, ensuring that it continues to run smoothly or to achieve specific outcomes.

This view also suggests that quantum entanglement isn't just a tool for communication or manipulation—it could be the very **nervous system** of the universe, allowing for instantaneous feedback between the operators and the simulation. Every quantum event, every entangled particle, contributes to this feedback loop, making the universe a constantly evolving, responsive system. The **laws of physics** we observe, while seemingly fixed, could be **dynamic**, responding to the needs and intentions of the operators as they guide the simulation toward its ultimate goal.

Quantum Entanglement and the Illusion of Free Will.

This also raises a profound question about the nature of **free will** within the simulation. If quantum entanglement allows the operators to monitor every aspect of the simulation in real-time, does this mean that our actions are predetermined? Or is free will an **illusion**, carefully crafted by the operators to maintain the integrity of the simulation?

From the perspective of the operators, **free will** might simply be another parameter—something to observe, measure, and manipulate as needed. Our choices, while seemingly independent, could be influenced by subtle **quantum interactions** that we are unaware of. The operators, using quantum engagement, might introduce small adjustments to the simulation, influencing the course of events without directly intervening.

For example, consider a scenario where a civilization is on the verge of discovering a critical technology—perhaps the ability to manipulate quantum systems on a large scale, thus threatening the stability of the simulation. The operators could use entanglement to nudge the outcomes, preventing certain discoveries or redirecting the course of progress without the inhabitants of the simulation ever realising it.

This idea introduces a delicate balance: while the simulation may **appear** to allow for free will, the operators could be subtly guiding its trajectory, ensuring that certain outcomes are achieved or avoided. **Quantum entanglement**, in this context, becomes the tool by which the operators maintain control, subtly shaping the flow of events without violating the illusion of autonomy that exists within the simulation.

However, there's another possibility to consider—perhaps the operators, despite their vast control over the simulation, can't interact directly with every person, creature, or event. This limitation might be similar to our relationship with the artificial intelligences we've created. While we design and train machine learning systems, we often lack full comprehension of what happens 'under the hood.' We can only control the output and attempt to align these systems with our goals, particularly concerning safety and alignment. In a similar vein, the operators may not be able to micromanage every detail of the simulation. Instead, they might rely on overarching rules and probabilistic adjustments to guide the overall trajectory.

Just as we impose safety measures and ethical frameworks on AI to ensure it behaves within acceptable bounds, the operators might be using the laws of physics and quantum entanglement as a form of alignment for the simulation. These constraints ensure the stability of the system without requiring direct intervention in every decision or event. From this perspective, the illusion of free will could be seen as a necessary component to maintain the integrity and autonomy of the simulation's inhabitants, while still enabling the operators to steer the simulation toward their intended outcomes.

In essence, the operators could be like creators of a complex, self-regulating AI system—able to monitor and guide, but not omnipotent in their influence. They might not have complete control over the thoughts and actions of every individual, just as we cannot fully predict the behaviour of our own AI creations. Instead, they observe the larger patterns, making subtle adjustments when needed, using mechanisms like quantum entanglement to nudge the simulation's course without ever being noticed. This raises profound questions about the nature of control and autonomy within the simulation, leaving us to ponder whether the operators are guardians of our perceived free will or simply caretakers of a system too complex even for them to fully comprehend.

The Quantum Backdoor and Higher Dimensions. Another possibility worth exploring is whether **quantum entanglement serves as a gateway** to **higher**

dimensions—the dimensions predicted by string theory that we cannot directly perceive. These dimensions, which exist beyond the familiar three of space and one of time, might be the true realm of the operators. In other words, the operators might exist in or have access to these **higher-dimensional spaces**, using quantum systems to interact with the lower-dimensional simulation we inhabit.

In this sense, entangled particles might be like **threads** connecting our universe to these higher dimensions. Just as a two-dimensional being might perceive a three-dimensional object as a series of shapes appearing and disappearing in its world, we might be witnessing **shadows of higher-dimensional phenomena** when we observe quantum entanglement. The operators, existing in these higher dimensions, could manipulate these threads, influencing the simulation from their vantage point.

This idea expands the role of **quantum mechanics** within the simulation hypothesis, suggesting that it isn't just a phenomenon that happens in our universe—it's a **cross-dimensional system** that allows for interaction between different layers of reality. The operators, from their higher-dimensional space, could pull on the strings of quantum entanglement, using it to influence events across multiple dimensions.

Quantum Entanglement and the Question of Consciousness.

Perhaps one of the most profound questions raised by this exploration is whether **consciousness itself** is tied to quantum systems. Could our ability to perceive, think, and make decisions be influenced—or even **created**—by quantum entanglement? If the operators are using quantum systems to monitor and guide the simulation, is it possible that they also **interface with our consciousness** through these same mechanisms?

There are theories in modern physics, such as **orchestrated objective reduction (Orch-OR)** proposed by physicist Roger Penrose and anaesthesiologist Stuart Hameroff, which suggest that consciousness might arise from **quantum processes** within the brain. This idea could be expanded within the context of your simulation hypothesis: if consciousness is indeed a quantum phenomenon, then perhaps the operators are able to **interact directly with our minds**, influencing thoughts, perceptions, and decisions through entanglement.

In this scenario, our minds are not just biological machines—they are **quantum interfaces** connected to the deeper fabric of the simulation. The operators, through quantum engagement, could influence not just the physical world, but also the **mental and emotional states** of those within the simulation. This raises profound questions about the nature of **free will, individuality, and**

reality—questions that blur the line between the physical and the metaphysical, the scientific and the philosophical.

The Operators' Endgame: Quantum Revelation?

Finally, as we dive deeper into the idea of quantum systems as the operators' tools, we must confront the possibility of an ultimate revelation—a moment when the simulation reaches its conclusion, and the true nature of reality is revealed. If quantum entanglement is the backdoor into the simulation, then perhaps there is a point at which this backdoor opens fully, allowing the inhabitants of the simulation to see beyond the veil.

In this final stage, the operators might use quantum entanglement to collapse the simulation, drawing all information back to themselves in an instant. Every quantum particle, every entangled system, could be part of a vast information retrieval system, designed to bring the simulation to its ultimate conclusion. The endgame of the simulation might be a moment of quantum revelation, when the entire universe is folded back into the higher-dimensional space from which it was born.

For the inhabitants, this moment could be one of transcendence or existential dissolution—a final confrontation with the ultimate question: What lies beyond the simulation? Would it be a higher state of being, a return to their true nature, or the realization that their existence was but a fleeting experiment?

As the simulation reaches its denouement, the operators would have achieved their goal—whether that goal is to test certain theories, explore the development of intelligent life, or even to answer a question posed by a higher entity. The quantum systems that have governed the universe throughout its existence would then serve their final purpose, bringing the experiment to a close.

But perhaps, for those within the simulation, the true endgame is not merely an answer to the operators' question, but the beginning of a new understanding—one that transcends the boundaries of the simulation itself.

End of chapter 5.

Chapter 6: The Expansion of the Universe: Computational Analogies

Cosmic Expansion as a Sign of Increased Memory and Processing Power.

The universe has been expanding ever since the Big Bang, but what if this isn't just a random cosmic event? Imagine, for a moment, that the universe is like a giant, ever-growing computer simulation. What happens when a simulation gets more complex? It needs more memory, more processing power, and more resources to keep running smoothly. So, could the expansion of the universe be the simulation's way of scaling up to handle all this complexity? And who might be controlling this process? Enter the operators—the unseen architects of this vast, cosmic system.

AI Scaling Laws and the Universe's Expansion.

Think about how AI models work. As they get bigger, with more data and parameters, they get better at what they do. The same might be true for the universe. Picture it as a massive AI model that's constantly learning and evolving under the guidance of the operators. To keep up with all the new stars, galaxies, and potentially trillions of lifeforms, it needs to keep expanding, just like a neural network that grows to tackle more complicated tasks.

In this grand simulation, each new galaxy or lifeform could be viewed as a unique subroutine or algorithm, designed to explore different aspects of existence. Imagine an AI model running multiple experiments simultaneously, each one contributing to a deeper understanding of its overarching objective. Similarly, the operators could be using this cosmic expansion to run countless simulations, each lifeform and civilization serving as a distinct piece of the cosmic puzzle.

Dark Energy: The Simulation's Optimization Algorithm.

Now, let's talk about dark energy, the mysterious force that's making the universe expand even faster. What if dark energy is like an optimization algorithm in an AI model? In machine learning, algorithms like gradient descent help AI models learn more efficiently by fine-tuning their parameters. Could dark energy be doing something similar for the universe—making sure everything runs smoothly as it expands?

For the operators, dark energy might be a way to ensure that the simulation doesn't overextend itself. As more lifeforms evolve and as civilizations advance, the demand on the simulation's resources grows. The accelerated expansion driven by dark energy could be a safeguard, allowing the simulation to accommodate all this added complexity without crashing or reaching its limits. It's almost as if the universe has its own built-in system updates, like patches that improve software performance over time, orchestrated by the operators.

Expansion and Increasing Complexity: The Role of Lifeforms.

As the universe expands, it becomes more complex. New galaxies form, stars are born, and with them, the possibility for new forms of life. It's like adding new features to a video game—every update makes the game more interesting and immersive. But this isn't just about humans. If there are billions or trillions of lifeforms out there, each could be acting as a different algorithm within the simulation, exploring different pathways of evolution, intelligence, and even consciousness.

Each species might be contributing to the overall simulation in its unique way, testing different variables that the operators want to understand. Perhaps some civilizations are meant to develop advanced technologies, while others are focused on biological or ecological systems. Some might even be exploring forms of existence we can't even comprehend. The universe's expansion, then, is not just to accommodate human understanding but to support the growth and evolution of all these lifeforms, each adding a new dimension to the simulation.

The Role of Human Understanding and the Operators' Vision.

Humans have always been curious, seeking to understand the universe and our place within it. But what if our quest for knowledge is just one part of a much larger plan? What if we, along with countless other lifeforms, are feeding information back to the operators? Every time we make a discovery—whether it's unlocking the secrets of quantum mechanics or exploring deep space—we could be contributing to a grand experiment orchestrated by the operators.

Imagine that every civilization, every intelligent species, is part of this vast information-gathering process. Each one might be designed to explore different aspects of reality, contributing data back to the operators who are piecing together the ultimate picture. As we push the boundaries of our understanding, the simulation has to expand, not just for us, but for all the civilizations out there. We're all part of a cosmic research project, driven by the operators' curiosity or purpose, whatever that might be.

The operators might be observing how different civilizations handle technological progress, social dynamics, or even existential questions. Are we, as humans, meant to figure out the simulation itself? Or are we just one of many species, each playing its part in a grander scheme? And what happens when all this information is gathered? Is there an ultimate goal, a final revelation waiting for us all?

Entropy and Thermodynamics in a Simulated Environment.

The universe, with all its complexity and order, is governed by fundamental principles like entropy and thermodynamics. But what if these aren't just physical laws? What if they're rules encoded into the very fabric of the simulation, ensuring everything runs smoothly and predictably? In a simulated universe, entropy might not just be a measure of disorder—it could be a crucial part of how information is processed and dispersed throughout the system.

Entropy as Information Dispersal: A Necessary Feature.

Imagine the universe as a vast computational network, with every particle, star, and galaxy acting as a data point. In such a system, entropy could represent the spread and flow of information, much like the way data is distributed across servers in a massive cloud network. Just as a well-designed computer system needs to prevent data bottlenecks and ensure efficient data flow, the universe needs to manage the distribution of energy and matter to maintain balance and stability.

In this context, the second law of thermodynamics—that entropy always increases—might not just be a rule of nature but a necessary feature of the simulation. It's like a built-in mechanism that ensures the simulation doesn't become too orderly or too chaotic, keeping everything within a functional range. This increase in entropy could be the universe's way of "balancing the books," ensuring that information is evenly distributed and accessible, just like how an AI model disperses data across different layers and nodes to process it efficiently.

AI Regularization and the Second Law of Thermodynamics.

In machine learning, techniques like regularization are used to prevent AI models from overfitting, ensuring they generalize well to new data. The second law of thermodynamics could be seen as a form of regularization for the universe. By pushing the system towards greater entropy, it prevents the simulation from becoming too predictable or too orderly, much like how regularization keeps an AI model flexible and adaptable.

Imagine the universe as a gigantic neural network. If everything were perfectly orderly, like a model that's memorized its training data, it wouldn't be able to adapt to new inputs or changes. The constant increase in entropy might be a way to introduce controlled randomness, keeping the universe dynamic and capable of evolving. It's as if the operators have set this rule to ensure the simulation remains robust and resilient, capable of handling whatever scenarios unfold within it.

Entropy and the Operators' Experiment.

From the perspective of the operators, entropy could serve a dual purpose. On the one hand, it ensures the stability and functionality of the simulation, but on the other, it could be a variable they're actively monitoring and manipulating. Different regions of the universe, with their varying levels of entropy, might be designed to test different aspects of the simulation.

For instance, high-entropy areas, like the vast voids between galaxies, could be places where the simulation is allowed to run on "auto-pilot," requiring minimal intervention. In contrast, low-entropy areas, such as regions of intense star formation or planets with burgeoning life, could be hotspots of activity, where the operators are closely observing or even intervening to guide the development of civilizations.

Each civilization, with its unique history and technological advancements, might be an experiment in how information evolves under different conditions of entropy.

Do civilizations in high-entropy environments develop differently from those in low-entropy ones? How does the balance of order and chaos influence the trajectory of intelligent life? These are the kinds of questions the operators could be exploring as they fine-tune the simulation.

The Role of Lifeforms in Entropy Management.
Life itself, with its tendency to create order out of chaos, plays a fascinating role in the universe's entropy. Every organism, from the simplest bacteria to the most advanced civilization, is engaged in a constant struggle against entropy, creating pockets of order in an ever-disordering universe. In the context of the simulation, lifeforms might act as localized systems that temporarily defy the overall increase in entropy, like subroutines in a computer program designed to maintain or even restore balance.

Think about human technology, for example. Every innovation, from fire to quantum computers, is a way of harnessing and directing energy, creating localized reductions in entropy. What if this isn't just a natural consequence of evolution but a built-in feature of the simulation? Perhaps lifeforms like us, and countless others throughout the universe, are part of the operators' method for managing entropy—small, self-contained systems that explore how order and chaos can coexist.

And it's not just humans. There could be countless civilizations, each with its own approach to entropy management. Some might be more technologically advanced, capable of manipulating entropy on a cosmic scale, while others might be focused on biological or ecological systems, maintaining balance in their own unique ways. Every civilization, every lifeform, could be part of a grand experiment, exploring the interplay between order and disorder in the simulation.

Entropy as a Measure of the Simulation's Progress. Finally, what if the total entropy of the universe is a measure of the simulation's progress? As entropy increases, the simulation might be moving towards a predefined endpoint, a moment when all the variables have been tested, and all the data has been collected. This could be the operators' way of tracking how far the simulation has come and how much further it needs to go.

In this view, the eventual "heat death" of the universe, where entropy reaches its maximum and all processes cease, could be the ultimate conclusion of the simulation. A final state where all information has been processed, and the simulation no longer needs to run. What happens then? Do the operators end the experiment, or is there a new phase awaiting us—one that we, as inhabitants of the simulation, can't even begin to comprehend?

Computational Limits and Their Physical Manifestations.

Just as a computer simulation is bound by the limits of its hardware, the universe—if it's a simulation—is constrained by its own set of rules and boundaries. These physical laws and constants, like the speed of light or the Planck scale, aren't just arbitrary. They could represent the simulation's built-in parameters, ensuring everything runs within the system's capabilities. But why would the operators impose such constraints, and what might happen if we, or any other civilization, were to push these limits?

The Universe as a Finite Simulation: Boundaries of the Sandbox.

Imagine you're playing a video game. No matter how vast the world seems, there are always boundaries—an invisible wall that you can't go beyond, or a point where the game simply won't render any more detail. The universe might work in a similar way. The observable universe could be the "sandbox" in which all events unfold, bounded by constraints like the speed of light and the laws of thermodynamics. These limits are like the edges of our cosmic playground, keeping everything within the manageable scope of the simulation.

From the operators' perspective, these boundaries are essential for the stability of the simulation. Without them, the universe could spiral into chaos, with unpredictable phenomena that could destabilize the entire system. The physical constants and laws that govern reality might be carefully chosen parameters, ensuring that the simulation can handle all the information and processes it's designed to explore.

The Planck Scale: The Universe's Pixel Resolution.

Think of the Planck scale as the pixel resolution of the universe. Just like an image on a screen can only be resolved to a certain degree of detail, the universe might have a smallest possible unit of space and time beyond which nothing can be measured or observed. This is the Planck length—a minuscule scale where the very fabric of space and time breaks down, beyond which the simulation might not be able to render any more detail.

For the operators, this limit could be a way to manage the simulation's resources efficiently. By setting a minimum resolution, they avoid wasting computational power on unnecessary details. It's like a game developer deciding not to render individual grains of sand on a distant beach.

The Planck scale is the universe's way of saying, "This is as detailed as it gets." Any attempt to probe beyond this boundary might be futile, as there simply isn't any more data to be found—at least, not within the confines of our simulated reality.

Physical Limits as Programming Constraints.

Every simulation has rules. In a video game, these might be the physics engine that determines how objects move and interact. In the universe, these rules are the physical constants and laws that we've spent centuries trying to understand. But what if these laws are more like lines of code, programming constraints that the operators have set to ensure the simulation runs smoothly?

The speed of light, for instance, could be seen as a speed limit imposed to prevent information from traveling faster than the simulation can process it. If particles could move faster than light, it might create paradoxes or inconsistencies that the simulation can't handle. Similarly, the constants that govern the strength of fundamental forces—like gravity or electromagnetism—might be fine-tuned parameters that keep the universe in a delicate balance, preventing it from collapsing or flying apart.

In this context, the universe's physical laws aren't just descriptors of how reality works—they're part of the operating system, rules that maintain the simulation's integrity. The operators, like programmers, have designed these rules to ensure the simulation behaves in a predictable and coherent way, allowing them to study the outcomes of their experiment without interference from chaotic, unmanageable phenomena.

Pushing the Limits: What Happens When Civilizations Test the Boundaries?

Throughout history, humans have been driven to push boundaries, to see what lies beyond the known world. The same is likely true for other intelligent lifeforms, wherever they may be. But what happens when civilizations start to test the very limits of the simulation? What if they develop technologies that approach the speed of light or manipulate quantum states in ways that challenge the underlying rules of the universe?

The operators might have anticipated this. Just as a game developer sets barriers to prevent players from glitching through the world or accessing forbidden areas, the universe could have built-in safeguards to keep us within the boundaries. The cosmic speed limit, the impenetrable nature of black holes, and even the seeming randomness of quantum mechanics could be ways to prevent us from breaking through the simulation's constraints.

But what if a civilization finds a way to bypass these barriers? Could they glimpse the "code" behind the simulation, or even interact with the operators themselves? It's possible that such breakthroughs are part of the experiment, a test to see if intelligent life can transcend its programmed limits. Perhaps some civilizations have already done so, escaping the sandbox and becoming something beyond our comprehension—a possibility that both excites and terrifies those who ponder it.

Physical Constants as the Operators' "Settings"

Every simulation has settings, parameters that can be adjusted to change how the system behaves. The operators might have set the physical constants of our universe with a purpose in mind, like a programmer tweaking the difficulty settings of a game. The fine-tuning of these constants has puzzled scientists for decades, as even a slight change could make the universe uninhabitable.

For the operators, these settings could be experiments in themselves. What happens if you set the gravitational constant just so, or tweak the speed of light? Perhaps our universe is just one of many simulations, each with slightly different settings, running in parallel or sequentially. The operators might be exploring how different rules affect the emergence and evolution of life, testing the robustness of their creation.

This could mean that the universe we inhabit is the result of countless iterations, a version that has been fine-tuned for a specific purpose. What that purpose is remains a mystery, but the fact that we exist in such a balanced, life-permitting universe might not be a happy accident—it could be the outcome of the operators' meticulous design, a set of parameters chosen to explore specific questions or scenarios.

The Edge of the Sandbox: Are There Places We Can Never Go?

Finally, consider the concept of an edge to the universe—not just in a physical sense, but as a boundary beyond which the simulation can't go. There might be regions of the universe that are inaccessible, places where the rules break down or where no information can penetrate. These could be like the unrendered parts of a video game map, where the simulation simply doesn't extend.

If such places exist, they could represent the true limits of the simulation, beyond which the operators themselves might not have intended us to explore. They could be barriers to prevent us from reaching areas of the "code" that we're not meant to see, or simply the outermost edges of the sandbox, where the simulation's resources are stretched to their limits. What lies beyond? Perhaps not even the operators know, or maybe that's where they reside, watching us as we edge ever closer to the boundaries they've set.

Dark Energy as Software Updates Expanding the Simulation.

Dark energy is one of the most enigmatic forces in the universe. It's the mysterious driver behind the accelerated expansion of the cosmos, pushing galaxies apart at an ever-increasing rate. But what if dark energy isn't just a random phenomenon? What if it's a deliberate feature of the simulation, a kind of cosmic "software update" introduced by the operators to keep the universe running smoothly and to expand the simulation's capabilities?

Dark Energy: The Universe's Patch Notes.

Think about how software updates work in a computer system. They fix bugs, add new features, and sometimes even expand the capabilities of the program. Dark energy might be doing something similar for the universe. It's as if the operators have realized that the simulation needs to accommodate more data, more complexity, and more lifeforms, and so they've introduced dark energy to expand the cosmic "hard drive."

Imagine the universe as a simulation that's growing beyond its initial capacity. Perhaps, in the early stages of the universe, there was no need for rapid expansion. But as the simulation evolved, with more galaxies, stars, and lifeforms, the operators might have realized that the system needed more space to run effectively. Dark energy could be their solution—a way to create more "room" for the simulation to continue evolving without reaching its computational limits.

Expansion Mechanisms in AI Systems: Dynamic Resource Allocation.

In AI systems, especially those dealing with massive datasets or complex tasks, dynamic resource allocation is crucial. The system needs to allocate memory and processing power based on the demands of the task at hand. Dark energy might be the universe's version of dynamic resource allocation, expanding the simulation as needed to handle the increasing complexity of the cosmos.

For example, in neural networks, resources are dynamically allocated to process and learn from data more efficiently. Similarly, dark energy could be expanding the universe to manage the increased "computational load" caused by the emergence of more complex structures, like black holes, advanced civilizations, and perhaps even the interconnected web of consciousness across different species. It's as if the universe is being optimized in real-time, with dark energy acting as a sort of cosmic load balancer.

Dark Energy as a Safeguard Against Computational Overload.

As more galaxies form and more civilizations potentially arise, the amount of information in the universe grows exponentially. Without dark energy, the simulation could become overcrowded, like a computer running too many applications at once. The result? Lag, crashes, or perhaps even a catastrophic failure of the simulation itself.

The operators might have anticipated this and introduced dark energy as a safeguard. By accelerating the expansion of the universe, they prevent any one region from becoming too densely packed with information. This expansion could help distribute the computational load evenly across the simulation, ensuring that no part of the universe becomes a bottleneck that could compromise the stability of the entire system.

The Operators' Role in Implementing "Software Updates"

From the operators' perspective, dark energy might be just one of many "patches" they've introduced to the simulation over its lifetime. Imagine them monitoring the universe like developers watching a complex simulation play out on a supercomputer. As they observe new challenges or potential issues arising—whether it's the formation of supermassive black holes, the emergence of advanced civilizations, or even the need for new physical laws—they implement updates to keep the simulation running smoothly.

These updates might not be limited to dark energy. Other unexplained phenomena, like the sudden appearance of new physics beyond the Standard Model, could be other forms of cosmic patches, introduced as needed to refine the simulation. The operators could be using these updates to test different scenarios or to guide the evolution of the simulation in a specific direction.

The Implications of a Continuously Updated Universe.

If the universe is continuously updated, what does that mean for us and other potential lifeforms within the simulation? For one, it suggests that the rules of the simulation aren't set in stone—they can change as needed. This could explain why certain physical phenomena remain unexplained or why constants in the universe seem so finely tuned. The operators might be adjusting these parameters in response to how the simulation unfolds.

Moreover, a continuously updated universe raises the question of whether we, as inhabitants of the simulation, have any influence over these updates. Could our actions, discoveries, and technological advancements trigger new patches or even entirely new "versions" of the universe? If we are part of a grand experiment, then perhaps our progress is being monitored, and the simulation is being adjusted to see how we respond to new challenges or opportunities.

The Endgame of Expansion: When Does It Stop?

If dark energy is a software update that keeps expanding the universe, when does it stop? Does the universe keep expanding indefinitely, or is there an endpoint? Perhaps the operators have a final goal in mind, a point where the simulation has collected all the necessary data, and the expansion is no longer needed.

The endgame of expansion might coincide with the endgame of the simulation itself. When the universe reaches a point where it can no longer sustain meaningful growth—whether because all possible variables have been tested or because all lifeforms have reached their potential—the operators might decide to end the experiment. At that point, the expansion could cease, and the simulation could be "folded back" into its original state, as discussed earlier.

Alternatively, the universe might continue expanding until it reaches a new phase, one that we can't yet comprehend. This could be the beginning of a new experiment, a new simulation, or a transition to a different kind of existence altogether. Whatever the case, the continuous expansion driven by dark energy suggests that the operators have a long-term plan in mind, one that goes far beyond our current understanding

.

Philosophical Implications of an Expanding Simulation

As we contemplate the universe expanding like a sophisticated simulation, it's natural to wonder what this means for us and our place in the grand scheme of things. If the cosmos is growing in response to an ever-increasing computational load, what are the philosophical implications for life, consciousness, and the nature of reality itself? Are we mere observers in a cosmic experiment, or do we play a more active role in shaping this expansive reality?

Expansion as a Metaphor for Growth and Evolution.

The constant expansion of the universe could be seen as a metaphor for growth and evolution. Just as the cosmos stretches further into the void, incorporating new stars, galaxies, and potentially new lifeforms, so too do we as conscious beings expand our understanding, our knowledge, and our capacity for growth. This parallel between the physical expansion of the universe and the intellectual and spiritual expansion of its inhabitants suggests a deeper connection between the macro and micro scales of existence.

In this sense, the operators might not just be expanding the universe to accommodate more data—they could be nurturing the evolution of consciousness itself. Every new discovery, every technological breakthrough, and every philosophical insight might contribute to the unfolding of the simulation's purpose. The universe, then, isn't just a static stage on which life plays out; it's an active participant in the growth and evolution of its inhabitants.

Humanity's Role as Co-Creators in the Simulation.
If the universe is a simulation, then we, as conscious beings, might not be just passive participants but active co-creators. Every time we make a discovery—whether it's mapping the human genome, exploring the quantum realm, or sending probes into deep space—we're feeding new information into the simulation. Our curiosity and desire to explore could be integral to the operators' plan, driving the expansion and complexity of the universe.

But it's not just humans. If there are countless other civilizations out there, each exploring their own corner of the cosmos, then they too are co-creators, each adding their unique perspectives and discoveries to the simulation. Imagine a network of interconnected minds, all contributing to a grand tapestry of knowledge and understanding. Each lifeform, each civilization, could be a node in a vast, cosmic neural network, helping the simulation evolve in ways we can't yet comprehend.

This idea raises profound questions about free will and purpose. Are we acting of our own volition, or are we following a script laid out by the operators? If we are co-creators, then perhaps we have more agency than we realize. Our choices, our discoveries, and even our failures might all be part of a larger design, a collaborative effort to shape the simulation's future.

The Role of Other Lifeforms: A Multitude of Perspectives.

If we're not alone in the universe, then what role do other lifeforms play in the expanding simulation? Each species, each civilization, might be exploring different aspects of reality, testing different variables, and providing diverse perspectives on the nature of existence. Some might be more technologically advanced, manipulating the very fabric of space and time, while others could be deeply connected to their planetary ecosystems, exploring the boundaries of biological evolution.

In this scenario, every lifeform could be a distinct algorithm within the simulation, contributing its unique insights to the operators' experiment. Just as diverse AI models can be trained on different datasets to solve unique problems, so too might each civilization be designed to explore a different aspect of reality. The expansion of the universe, then, isn't just about accommodating more matter and energy—it's about accommodating more ways of understanding, more ways of being.

The operators could be observing how different civilizations handle challenges like scarcity, conflict, or even the realization that they are part of a simulation. How do different species react to the knowledge that their reality might be artificial? Do they strive to break free, to understand their creators, or do they accept their existence and continue to evolve within the constraints of the simulation? These are the kinds of questions that the operators might be exploring as they expand the universe to encompass more lifeforms and more possibilities.

The End of Expansion: Is There a Final Goal?

If the universe is expanding to incorporate more data and more experiences, then what happens when there's nothing left to explore? Is there an endgame to this expansion, a point at which the operators achieve their goal and the simulation is complete? Or does the universe continue to expand indefinitely, forever incorporating new forms of life, new experiences, and new ways of understanding?

One possibility is that the operators have a specific objective in mind—a final revelation, a point where all the data has been collected, and all the variables have been tested. At that moment, the simulation might reach a state of perfect knowledge, a kind of cosmic singularity where everything is understood, and nothing more can be added. This could be the ultimate purpose of the simulation, a moment of cosmic enlightenment where the boundaries between the simulated and the real dissolve, and the inhabitants of the universe finally see beyond the veil.

Alternatively, the expansion could be endless, an infinite game where the purpose is not to reach an endpoint but to keep exploring, to keep evolving. In this view, the operators might not be looking for a final answer but are instead fascinated by the process of exploration itself. The universe, then, is a playground of infinite possibilities, a stage where the drama of existence plays out in countless forms, each one contributing to the ongoing story of the cosmos.

Our Place in the Expanding Simulation: Meaning and Purpose.

What does all this mean for us, here and now? If we are part of an expanding simulation, then our lives, our choices, and our experiences are all part of a larger story, a story that's still being written. This realization can be both exhilarating and humbling. It suggests that we are more than just biological entities struggling for survival—we are participants in a grand experiment, co-creators of a reality that's far more complex and beautiful than we can imagine.

But it also raises questions about purpose and meaning. If we are part of a simulation, then what is our purpose? Is it simply to live and learn, to love and explore, or is there something more? Perhaps the answer lies in how we choose to engage with the simulation. If we see ourselves as passive observers, then our role is limited. But if we embrace our role as active participants, as co-creators, then the possibilities are endless.

In the end, the expansion of the universe might be a call to action, an invitation to engage with the mystery of existence, to push the boundaries of our understanding, and to contribute our unique voice to the chorus of life.

Whether the universe has a final goal or is an infinite playground, our role is to explore, to create, and to be fully present in this cosmic dance.

End of chapter 6.

Chapter 7: Addressing the Fermi Paradox and Extraterrestrial Life

Revisiting the Fermi Paradox in the Context of the Simulation

The Fermi Paradox is traditionally framed as a contradiction between the high probability of extraterrestrial life and the lack of evidence for it. In a simulated universe, however, this contradiction might not be a mystery at all. It could be a feature deliberately designed by the operators. Just as game developers create boundaries and scenarios to guide players through a narrative, the operators might be using the Fermi Paradox as a way to keep civilizations isolated, ensuring that they follow certain paths of development without interference from each other.

Imagine a video game where players explore a vast, open world, but there are certain areas they cannot access until they've reached a specific level or completed certain tasks. Similarly, the operators might be controlling our access to other civilizations, only allowing contact when specific conditions are met. This could explain why, despite the enormous number of potentially habitable planets, we have yet to find any sign of intelligent life. We're simply not "ready" to access those parts of the simulation.

Time Lags, Communication Barriers, and the Simulation's Constraints.

Even if other civilizations exist within the simulation, the laws of physics—particularly the speed of light—create formidable barriers to communication. Signals from distant civilizations would take thousands, millions, or even billions of years to reach us. By the time we receive them, those civilizations might be long gone, or they could have evolved into forms of life or technology beyond our comprehension.

But what if these barriers aren't just natural consequences of the universe's physical laws? What if they're part of the simulation's design, intentional constraints imposed by the operators to prevent us from establishing contact? The speed of light, for example, could act as a "firewall" in the simulation, a way to keep information from traveling too quickly, thus maintaining the isolation of different civilizations. This would allow each civilization to develop independently, without the risk of premature contact that could disrupt the simulation's intended scenarios.

Think of the speed of light as a cosmic "parental control," limiting our access to other parts of the universe until we've matured as a species. The operators might be watching to see how we handle our isolation, whether we destroy ourselves, or whether we evolve to a point where we're capable of responsible interstellar communication. The time lags and communication barriers could be tests, measuring our readiness to join a larger cosmic community.

The Drake Equation and Probability of Life Within the Simulation.

The Drake Equation, developed to estimate the number of active, communicative extraterrestrial civilizations in the Milky Way, takes on a new meaning in the context of the simulation hypothesis. Traditionally, the equation multiplies a series of factors—such as the rate of star formation and the fraction of planets that could support life—to arrive at a probability of intelligent civilizations. But what if these factors are variables in the simulation, deliberately set by the operators?

The probability of life might not be a natural outcome but a controlled parameter. The operators could be adjusting the variables in the equation to create specific scenarios. For example, they might want to see how a single, isolated civilization (like humanity) develops under the assumption that it's alone in the universe. Alternatively, they could be running parallel simulations with different values for the Drake Equation, testing how different levels of interstellar interaction affect the development of civilizations.

In this view, the Drake Equation isn't just a tool for estimating the number of extraterrestrial civilizations—it's a reflection of the operators' choices, a glimpse into the settings of the simulation. Each term in the equation could correspond to a different experimental parameter, set by the operators to explore different outcomes. This perspective transforms the Drake Equation from a probabilistic formula into a coded message, hinting at the underlying structure of the simulation.

Simulation Boundaries: Limiting Interaction Between Civilizations.

If we are part of a simulation, there may be invisible boundaries limiting our ability to interact with other civilizations. These boundaries might not be physical walls but constraints built into the simulation's rules, preventing us from reaching certain parts of the universe or developing technologies that would allow us to make contact. But why would the operators impose such constraints, and what might their purpose be?

Cosmic "Sandbox Mode": Keeping Civilizations in Isolation.

Imagine the universe as a giant sandbox game, where different civilizations are like players building and exploring their own worlds. In a sandbox game, players are often kept in separate areas to develop their skills and strategies before they're allowed to interact with others.

The same principle might apply to civilizations within the simulation. The operators could be keeping us in "sandbox mode," ensuring that we develop independently before we're introduced to the larger cosmic community.

This isolation could serve multiple purposes. It might prevent less advanced civilizations from being overwhelmed or disrupted by more advanced ones. It could also allow each civilization to develop its unique culture, technology, and understanding of the universe without interference. By keeping civilizations separate, the operators ensure a diverse range of outcomes, each providing valuable data for their experiment.

But this raises an intriguing question: Are there civilizations out there that have graduated from this "sandbox mode"? If so, what did they achieve to earn that status, and what awaits them beyond the boundaries of their isolated worlds? Perhaps the operators are waiting to see if we, too, can reach that point—if we can demonstrate the wisdom, resilience, and ingenuity necessary to transcend our isolation and join the greater simulation.

The Cosmic Firewall: Preventing Premature Contact.

The speed of light, vast distances between stars, and the sheer size of the universe act as natural barriers to contact between civilizations. But what if these barriers are more than just physical constraints? What if they're part of a deliberate design—a "cosmic firewall" meant to prevent premature contact?

In a simulation, the operators might use these barriers to control the flow of information, ensuring that civilizations only interact when the time is right. Just as a firewall in a computer system protects sensitive data and controls network access, the cosmic firewall could be protecting the integrity of the simulation, preventing disruptions that could arise from uncontrolled interaction between civilizations.

This might explain why, despite the vastness of the universe and the high probability of extraterrestrial life, we have yet to encounter any sign of it. The firewall keeps us in a kind of quarantine, allowing us to develop independently without external influence. It's a way of ensuring that each civilization's development follows its unique trajectory, unhindered by the actions or knowledge of others.

But what happens when a civilization finds a way to bypass this firewall? Could they become a threat to the stability of the simulation, or would they be welcomed as pioneers who have earned the right to explore beyond the boundaries? The operators might be watching closely, ready to intervene if necessary, or perhaps even encouraging such breakthroughs as part of their grand experiment.

Psychological and Cultural Barriers: Subtle Forms of Isolation.

Not all barriers are physical. Some are psychological or cultural, shaping our beliefs, desires, and motivations. These subtle forms of isolation could be just as effective in preventing contact with other civilizations. Imagine if the operators have subtly influenced our collective consciousness, steering us away from the pursuit of interstellar contact.

This could manifest in various ways. Perhaps our fascination with local space exploration, like missions to Mars and the Moon, is part of this influence, keeping us focused on nearby goals rather than looking outward. Or consider the many cultural and philosophical barriers that make the idea of contacting extraterrestrial life seem distant or improbable. These could be subtle nudges from the operators, guiding us along a path that keeps us isolated.

If this is the case, then breaking free from this isolation might require more than just technological advancements. It could mean overcoming deep-seated beliefs and cultural narratives that limit our vision of what's possible. It could involve a shift in our collective mindset, a new way of thinking that embraces the possibility of contact and actively seeks to transcend the barriers, both seen and unseen, that keep us isolated.

Simulation Boundaries as Protective Measures.

Another possibility is that the simulation boundaries are protective measures, designed to shield us from dangers that we're not yet equipped to handle. Just as parents keep children from exploring dangerous environments until they're old enough to understand the risks, the operators might be keeping us from encountering more advanced or even hostile civilizations.

This perspective suggests that the universe is a much more complex and potentially dangerous place than we realize. There could be forces or entities out there that are beyond our comprehension, capable of destabilizing not just our civilization but the entire simulation. The operators, in this scenario, are acting as guardians, ensuring that we are kept safe within the confines of our isolated corner of the cosmos.

This raises profound questions about what lies beyond the boundaries. What are we being protected from? Are there cosmic predators, rogue civilizations, or even glitches in the simulation that we're not meant to encounter? If we are being kept in a kind of cosmic nursery, then what will it take for us to prove that we're ready to leave and face the unknown?

Testing the Boundaries: Are We Meant to Push Them?

If the boundaries are part of the simulation, then are we meant to test them? History is full of examples of humans pushing against the limits of their environment, from crossing oceans to exploring space. This drive to explore, to push boundaries, might be a fundamental part of who we are as a species. But what if it's also part of the operators' experiment?

Perhaps the operators have designed the simulation to see if we, and other civilizations, have the curiosity and determination to push beyond our constraints. The barriers are there to be challenged, to see if we can find creative solutions to transcend them. This could be a test of our ingenuity, resilience, and desire for knowledge—a way to see if we're worthy of advancing to the next level of the

simulation.

If this is the case, then our efforts to explore the universe, to develop new technologies, and to expand our understanding of reality are more than just scientific endeavours. They're part of a cosmic test, a challenge issued by the operators to see if we can break free from our isolation and discover the true nature of the universe. It's a test that we may not even be aware of, but one that could determine our ultimate destiny.

The Role of Advanced Civilizations: Guardians, Guides, or Prisoners?

If there are more advanced civilizations out there, what role do they play in the simulation? Are they guardians, working with the operators to protect less advanced civilizations like ours? Or are they guides, waiting to welcome us once we've proven ourselves worthy of joining the larger cosmic community?

Another possibility is that they, too, are prisoners of the simulation, confined within their own boundaries, unable to interact with others until they've met certain criteria. If that's the case, then the universe might be filled with isolated civilizations, each struggling to break free from their constraints, each searching for the operators and the true nature of their reality.

The idea of advanced civilizations as fellow prisoners adds a layer of complexity to the Fermi Paradox. It suggests that we are not alone in our isolation, that there are others out there facing the same challenges, the same mysteries. Perhaps the operators are watching to see which civilizations can find a way to communicate, to cooperate, to break through the barriers and discover the true nature of their existence.

Expanding Consciousness: A New Form of Communication.

Technology isn't the only pathway to overcoming isolation. We might also need to expand our consciousness, developing new ways of thinking and perceiving that allow us to connect with other forms of life—or even the operators—on a deeper level. Throughout history, humans have sought altered states of consciousness through meditation, trance, and the use of psychoactive substances. Among these, one substance has gained particular notoriety for its profound and consistent effects: DMT, or dimethyltryptamine.

Often referred to as the "spirit molecule," DMT induces experiences that many describe as contact with otherworldly beings or entities. These encounters often include a sense of being in the presence of an intelligence far greater than one's own, and a feeling of receiving information or guidance that transcends ordinary reality. While these experiences are subjective and difficult to interpret scientifically, they raise fascinating questions about the nature of consciousness and the potential for communication beyond the physical realm.

What if DMT and similar substances are not just chemical triggers for hallucinations, but gateways to a deeper level of reality? Taking DMT could be seen as akin to consuming a "power-up" in a video game—a temporary enhancement that allows players to access hidden levels, unlock new abilities, or gain insights that are otherwise inaccessible. For those brief moments, individuals might be tuning into a frequency outside the usual parameters of the simulation, accessing a level of reality that is normally hidden from us.

This could be a deliberate feature of the simulation, a kind of "hidden menu" that allows for communication between the inhabitants of the simulation and the operators. If true, then the implications are staggering. It would mean that the operators have built in a way for us to reach out to them, to ask questions, and perhaps even to receive answers. But this method of communication is not easy to access; it requires a radical shift in consciousness, one that most people are unprepared for.

The experiences reported by those who have used DMT and similar substances could be seen as glimpses into the operators' domain, a fleeting connection to the minds behind the simulation. This would explain the profound, often life-changing nature of these experiences, and the difficulty in articulating them afterward. If these substances allow us to briefly step outside the simulation, then the challenge is not just in accessing this state, but in bringing back meaningful information that can be integrated into our understanding of reality.

The Ethical and Philosophical Implications of Using Altered States for Communication.

If DMT and other substances do allow for communication with the operators, then this raises profound ethical and philosophical questions. Should we seek to use these tools to contact the operators, or is this a form of trespassing, an intrusion into a realm we are not meant to enter? And if the operators respond, what would that mean for our understanding of free will, autonomy, and the nature of reality itself?

There is also the question of interpretation. How can we be sure that what is experienced under the influence of these substances is a genuine encounter with the operators, and not just the mind's way of processing an overwhelming influx of information? The subjective nature of these experiences makes them difficult to study, and even more difficult to interpret. If we are to use altered states of consciousness as a tool for communication, we would need to develop a framework for understanding and integrating these experiences, one that respects both the potential insights and the limitations of this approach.

Expanding the Simulation: A Purposeful Expansion of Consciousness?

If the operators have included this potential for expanded consciousness within the simulation, then perhaps they are not just watching us but encouraging us to seek them out. The existence of these substances in nature, and the profound experiences they induce, could be seen as a kind of invitation—a challenge to expand our minds and to explore the boundaries of what it means to be conscious.

In this view, the simulation is not just a test of our technological and scientific capabilities but of our spiritual and philosophical growth as well. The operators might be waiting to see if we can transcend the materialistic, reductionist view of reality that dominates our current understanding and embrace a more holistic, integrated perspective. This would mean seeing consciousness not as a byproduct of matter, but as a fundamental aspect of the universe, one that connects us to the operators and to each other in ways we are only beginning to understand.

Conclusion: The Role of Expanded Consciousness in Overcoming Isolation.

The idea that altered states of consciousness could provide a means of communication with the operators is a speculative one, but it resonates with the themes of this book. It suggests that the boundaries of the simulation are not just physical or technological but also psychological and spiritual. Overcoming these boundaries might require us to explore not just the outer reaches of space, but the inner reaches of our own minds.

If we can learn to navigate these altered states of consciousness, to use them as tools for exploration and communication, then we might find that the answers we seek are closer than we think. The operators might not be distant, unreachable entities, but beings who have been waiting for us to reach out, to make contact in the only way that truly transcends the limitations of the simulation—through the infinite, boundless potential of consciousness itself.

The Operators' Purpose: Testing Isolation and Interaction

The operators, as the architects of our simulated reality, may have many reasons for keeping civilizations isolated. It could be to see how different forms of life evolve and adapt under the constraints of solitude, or it could be to prevent disruptive interactions that might destabilize the simulation. But if their goal is to understand how intelligent beings respond to their isolation, what are they looking for? And what might happen when we, as a civilization, demonstrate that we are ready to break free from our imposed boundaries?

Isolation as a Catalyst for Innovation and Growth. One possibility is that isolation is intended to foster innovation and growth. When a civilization is left to its own devices, without the influence or interference of more advanced beings, it is forced to rely on its own creativity and ingenuity to solve problems and overcome challenges. This could be a deliberate strategy by the operators to encourage the development of unique cultures, technologies, and philosophies—an experiment in the diversity of intelligent life.

Just as a plant grows stronger in the absence of artificial support, a civilization might become more resilient and resourceful when it believes it is alone in the universe. The operators could be watching to see which civilizations are able to rise above their circumstances, to create, explore, and innovate despite their isolation. Those that succeed might be rewarded with contact, while those that fail might be left to continue their solitary existence.

This would suggest that the Fermi Paradox is not a paradox at all, but a test—a way for the operators to gauge the readiness of a civilization for the next stage of development. It's not just about finding intelligent life; it's about proving that we are capable of thriving and growing in the face of adversity, of pushing beyond our limitations to achieve something truly remarkable.

Interaction as a Catalyst for Understanding and Unity.

While isolation can drive innovation, interaction might be the key to a deeper understanding of the universe and our place within it. If the operators are testing how civilizations respond to isolation, they might also be interested in how we handle the prospect of contact. What happens when we, or another civilization, finally break through the barriers that have kept us apart? How do we react to the discovery that we are not alone?

This could be the ultimate test of our maturity as a civilization. Are we able to embrace the unknown, to accept the existence of beings who are different from us, and to seek understanding and cooperation rather than conflict and domination? Or do we react with fear, hostility, and a desire to protect our own interests at the expense of others?

The operators might be watching to see if we are capable of reaching out in peace, of forming alliances and partnerships that transcend the boundaries of our isolated worlds. If we can demonstrate this level of understanding and unity, then we might be ready for the next stage of the simulation—one in which we are no longer confined to our own corner of the universe but are free to explore and interact with others.

The Endgame: What Happens When the Test is Complete?

If the Fermi Paradox is a test, then what happens when the test is complete? When a civilization demonstrates that it is ready to move beyond its isolation, does the simulation change? Are the barriers lifted, allowing

for contact with other civilizations and perhaps even with the operators themselves? Or is there a more radical transformation, one that alters the very nature of the simulation?

One possibility is that the simulation is not a static system but a dynamic one, capable of evolving and adapting in response to the actions of its inhabitants. When a civilization reaches a certain level of development, the simulation might undergo a kind of "upgrade," expanding to include new possibilities and new challenges. This could involve the introduction of new physical laws, the appearance of new forms of life, or even the opening of new dimensions of reality.

Alternatively, the simulation might come to an end, having achieved its purpose. The operators could decide that there is nothing more to learn from this particular scenario and shut it down, either to start anew or to move on to other experiments. This raises the unsettling question of what happens to the inhabitants of the simulation when it ends. Do we cease to exist, or do we transition to a new state of being, one that transcends the boundaries of the simulated universe?

Transcending the Simulation: The Ultimate Goal?

If the ultimate goal of the operators is to see if civilizations can transcend the simulation itself, then the Fermi Paradox takes on an even deeper significance. It's not just about isolation and interaction but about the potential for a civilization to reach a level of understanding that allows it to see beyond the simulation, to understand the nature of its existence and to break free from the constraints of its reality.

This could be the ultimate test: to see if we can recognize the limitations of our universe and find a way to transcend them. It might involve developing new forms of science and technology that go beyond our current understanding, or it could mean achieving a new level of consciousness that allows us to perceive and interact with the operators directly.

If this is the case, then the Fermi Paradox is not a problem to be solved but a mystery to be embraced. It is a challenge issued by the operators, a call to explore the limits of what is possible and to push beyond them. It is an invitation to discover the true nature of reality, to find the hidden pathways that lead out of the simulation and into the unknown.

Preparing for Contact: The Role of Culture and Imagination.

If we are to overcome our isolation and prepare for contact with other civilizations—or even with the operators—then culture and imagination will play a crucial role. We need to expand our understanding of what is possible, to open our minds to new ideas and new ways of thinking. This means embracing not just science and technology but art, philosophy, and spirituality as well.

Our myths, stories, and dreams are not just idle fantasies; they are explorations of the boundaries of reality, attempts to reach beyond the limits of our understanding and to glimpse what lies beyond. They are the seeds of the future, the visions that inspire us to create and to explore. If we are to prepare for contact, we need to cultivate our imagination, to embrace the unknown and to be willing to explore the uncharted territories of the mind and spirit.

This could be why the operators have left the Fermi Paradox unresolved. They are not just testing our technological prowess but our capacity for wonder, curiosity, and creativity. They want to see if we can imagine a reality beyond our own, if we can dream of possibilities that transcend the boundaries of the simulation. If we can, then we might be ready to meet the unknown, to face the mystery of our existence, and to discover the true nature of the universe.

The Operators' Legacy: What Will We Leave Behind?

Finally, as we contemplate the Fermi Paradox and the possibility of overcoming our isolation, we must consider the legacy we will leave behind. If we are being watched, studied, and tested, then what will future generations—whether human, extraterrestrial, or something else—learn from us? What stories will they tell about our time in the simulation, about our struggles and triumphs, our fears and hopes?

Our legacy is not just what we achieve technologically or scientifically but how we live, how we treat each other, and how we face the challenges of existence. It is the story we tell about ourselves, a story that will be passed down through the ages, a testament to our humanity and our capacity for growth and change.

In the end, the Fermi Paradox is not just a question about extraterrestrial life; it is a question about who we are and who we want to be. It is a mirror that reflects our deepest fears and our highest aspirations, a challenge that calls us to reach beyond our limitations and to discover the true potential of our existence. Whether we are alone or not, whether we ever make contact or not, the journey itself is what matters. It is the journey that defines us, that gives our lives meaning, and that will determine the legacy we leave behind.

End of chapter 7

Chapter 8: Philosophical and Ethical Considerations

Nick Bostrom's Simulation Argument: Implications and Expansions

Nick Bostrom's Simulation Argument has sparked widespread intrigue and debate since it was first proposed in 2003. At its core, Bostrom presents a provocative idea: one of three possibilities must be true. Either no civilization will ever reach the technological maturity to create simulated realities, or advanced civilizations that do reach such a level will choose not to create simulations, or—and here is where things get truly interesting—we are almost certainly living in a simulation.

If we accept this third proposition, then everything we know about reality could be an illusion, a construct designed by some higher intelligence or, as we've framed it in this book, the Operators. But Bostrom's argument, while compelling, leaves many questions unanswered. Who are these Operators? What might be their purpose in creating and maintaining these simulated realities? And what does it mean for us, as simulated beings, to grapple with the implications of this possibility?

To delve deeper into Bostrom's argument, we first need to revisit its foundations. Bostrom's theory hinges on the idea that if any advanced civilization were to develop the capability to create highly sophisticated simulations, they would likely do so many times over. The number of simulated realities would, therefore, vastly outnumber the single "base" reality. The result? The odds that we are living in the base reality, rather than one of the countless simulations, are infinitesimally small.

So, what does this mean for us? If we are indeed living in a simulation, it fundamentally alters our understanding of existence. Our lives, our experiences, even our thoughts could all be part of a complex program designed by entities far beyond our comprehension. This idea, while unsettling, also opens up a wealth of possibilities. For one, it suggests that our reality might be one of many—a single simulation within a vast multiverse of simulations.

Expanding the Argument: The Role of the Operators.

Bostrom's original argument doesn't delve into the nature of the beings who might create these simulations. For our purposes, we refer to them as the Operators. If the Operators exist, they are not just passive observers but active participants in shaping the course of the simulated universe. This perspective introduces a layer of intentionality to the simulation hypothesis that Bostrom's framework doesn't fully explore.

Why would the Operators create a simulation like ours? One possibility is that they are conducting experiments, exploring how intelligent beings develop under different conditions. Perhaps they are testing various philosophical or ethical theories, using simulations as a controlled environment to observe the outcomes. Another possibility is that the simulation serves as a kind of entertainment or educational tool for the Operators, much like how we create video games or historical simulations.

Expanding on Bostrom's idea, we can also imagine the possibility of multiple nested simulations. In this scenario, the Operators themselves could be simulated beings, created by even higher-level Operators. This recursive structure of nested realities would create a multiverse of simulations, each with its own set of rules, purposes, and inhabitants. In this view, our universe is just one layer in an infinitely complex hierarchy of simulations.

Implications for Reality and Existence.

Accepting the possibility that we are living in a simulation challenges our most fundamental assumptions about reality. If everything we perceive is part of a programmed environment, what does that say about the nature of our experiences? Are our emotions, our struggles, our triumphs any less real because they occur within a simulated framework?

From a psychological perspective, the simulation hypothesis can be both disorienting and liberating. On one hand, it might lead to a sense of existential detachment, a feeling that nothing truly matters if it's all just part of a grand illusion. On the other hand, it could inspire a sense of wonder and curiosity. If we are in a simulation, then what lies beyond it? Who are the Operators, and what are their motivations? These questions can drive us to explore the boundaries of our understanding, pushing us to seek answers that transcend the limitations of our perceived reality.

Moreover, if we are living in a simulation, what does that mean for our sense of self and consciousness? Are we truly conscious, or are we simply highly sophisticated programs designed to mimic consciousness? This question touches on the age-old mind-body problem, challenging our understanding of what it means to be self-aware. If consciousness is just a byproduct of the simulation, then it raises the possibility that our identities, our very sense of being, are constructs no different from the pixels on a screen or the code in a software program.

Ethical and Moral Implications.

The ethical dimensions of the simulation hypothesis are equally profound. If the Operators have control over every aspect of our reality, do they bear moral responsibility for the suffering and challenges within it? Are they like deities in a traditional religious framework, omnipotent and omniscient beings who can intervene at will? Or are they more like scientists, observing their experiment from a distance, refraining from interference to maintain the integrity of their research?

If we accept that the Operators have the power to alter the simulation, then why do they allow suffering and injustice to exist? This question, often referred to as the problem of evil, has been a central concern in theology and philosophy for centuries. In the context of the simulation hypothesis, it takes on a new dimension. Are we simply players in a game, our struggles serving some greater purpose that we cannot fathom? Or are we the subjects of a grand experiment, our pain and joy mere data points in a cosmic spreadsheet?

On a more personal level, the simulation hypothesis also forces us to reconsider our own ethical responsibilities. If we are living in a simulation, does that change how we should behave? Should we strive to live by the same moral principles, or does the knowledge that our reality is artificial render those principles irrelevant? These questions challenge us to think deeply about the nature of ethics and morality, pushing us to consider how we should act in a world where the lines between reality and illusion are blurred.

Conclusion: Bostrom's Argument as a Gateway to Deeper Questions

Bostrom's Simulation Argument serves as a gateway, inviting us to explore questions that go beyond the realm of traditional philosophy and science. By considering the possibility that we are living in a simulation, we are forced to confront the nature of reality, the essence of consciousness, and the foundations of morality. As we delve deeper into these questions, we find that they are not just abstract philosophical puzzles, but issues that touch on the very core of what it means to be human.

In the following sections, we will continue to explore these themes, drawing on the ideas of other philosophers and thinkers to further illuminate the ethical and existential dimensions of the simulation hypothesis. Through this journey, we aim to gain a deeper understanding of our place in the universe—whether it is real or simulated.

René Descartes and Scepticism of Reality: Parallels in the Simulation Hypothesis.

René Descartes, the 17th-century philosopher often regarded as the father of modern philosophy, grappled with questions about the nature of reality that remain strikingly relevant today, especially in the context of the simulation hypothesis. Descartes' approach to philosophy began with radical doubt, famously encapsulated in his method of doubting everything that could possibly be doubted. This process led him to one undeniable conclusion: "Cogito, ergo sum" ("I think, therefore I am"). In his quest for certainty, Descartes realized that even if he were being deceived by an all-powerful being, the very act of doubting proved his existence as a thinking entity.

The parallels between Descartes' scepticism and the simulation hypothesis are both profound and unsettling. Descartes imagined an "evil demon," a malicious entity that could manipulate his perceptions and deceive him into believing in a reality that did not exist. Today, we might replace Descartes' demon with our notion of the Operators—entities who have the power to create and control a simulated universe, deceiving us into believing in a reality that is, at its core, an illusion. The modern concept of a simulated reality is a high-tech version of Descartes' age-old philosophical conundrum: How can we be sure that what we perceive is real?

Descartes' Method of Doubt and the Simulation Hypothesis.

Descartes' method of doubt was a revolutionary approach to philosophy. By systematically doubting the reliability of his senses, the existence of the physical world, and even the truths of mathematics, he sought to strip away all beliefs that could be called into question. His goal was to find a foundation of certainty upon which all knowledge could be built. In this process, Descartes considered the possibility that everything he experienced might be an illusion created by a powerful deceiver.

The simulation hypothesis takes this line of thinking a step further. It suggests that our entire universe, not just our sensory experiences, could be an artificial construct. The Operators, like Descartes' evil demon, have the ability to manipulate everything we perceive, from the stars in the sky to the thoughts in our heads. If the Operators control the rules and parameters of our reality, then how can we ever be certain that what we experience is true?

Descartes ultimately concluded that his own existence as a thinking being was the only certainty he could find. Even if he was being deceived, the fact that he was capable of thought proved his existence. The simulation hypothesis, however, challenges even this conclusion. If our thoughts are merely lines of code in a vast computational system, then what does it mean to exist? Is self-awareness proof of our existence, or could it be a sophisticated illusion created by the Operators?

The Role of the Observer: Consciousness in a Simulated World.

Descartes' famous declaration, "I think, therefore I am," places consciousness at the centre of his philosophy. For Descartes, the act of thinking, doubting, and questioning was the essence of being. In the context of the simulation hypothesis, this raises a fascinating question: What role does consciousness play in a simulated reality?

If our reality is a simulation, then consciousness could be seen as a kind of anomaly, a phenomenon that cannot be fully explained by the code and algorithms governing the simulation. Alternatively, consciousness might be a fundamental part of the simulation, an emergent property designed by the Operators to explore complex scenarios and ethical dilemmas. In either case, the simulation hypothesis suggests that consciousness is more than just a byproduct of physical processes; it is a central feature of the simulated universe.

This perspective invites us to reconsider the nature of consciousness itself. Are we truly conscious beings, capable of independent thought and action, or are we simply sophisticated programs, following a script written by the Operators? Descartes believed that the mind and body were separate entities, a view known as dualism. The simulation hypothesis, with its suggestion that our physical bodies are simulated constructs, brings new relevance to this debate. If our bodies are virtual, does that mean our minds exist outside the simulation, or are they also part of the code?

Operators as the New "Evil Demon"

Descartes' evil demon was a being of unimaginable power, capable of deceiving him about the nature of reality. In our framework, the Operators serve a similar role. They are the creators and controllers of the simulation, able to manipulate every aspect of our perceived universe. However, while Descartes' demon was malevolent, the intentions of the Operators remain a mystery. Are they benevolent beings, guiding us towards enlightenment? Or are they indifferent scientists, observing their experiment without regard for the experiences of their simulated subjects?

This comparison raises ethical and existential questions. If the Operators are capable of intervening in our reality, do they have a moral obligation to do so? Should they protect us from harm, or are we simply data points in a grand experiment? These questions echo the concerns Descartes had about the nature of the evil demon, but with a modern twist that incorporates the technological possibilities of our age.

Moreover, if the Operators are deceiving us, even unintentionally, can we ever hope to see through their deception? Descartes believed that by using reason and doubt, he could achieve certainty. The simulation hypothesis, however, suggests that there may be no way to break free from the illusion. If the simulation is perfect, then even our doubts and questions could be part of the design. In this way, the Operators are not just deceivers; they are the architects of our reality, shaping the very way we think and perceive.

Certainty and Doubt in the Simulation.

Descartes' philosophy ultimately sought a foundation of certainty—something that could not be doubted. But in a simulated reality, can we ever achieve such certainty? If the Operators control not just our perceptions but the very logic and rules of our universe, then what can we trust? Even the laws of physics, which seem immutable, could be arbitrary rules set by the Operators. In this context, Descartes' quest for certainty seems both heroic and tragically doomed.

The simulation hypothesis suggests that doubt is not just a philosophical tool but a necessary response to the nature of our reality. If we cannot trust our senses, our logic, or even our own thoughts, then what can we rely on? This radical scepticism can be disorienting, but it also pushes us to think more deeply about the nature of existence. Perhaps the true value of the simulation hypothesis, like

Descartes' method of doubt, is not in the answers it provides but in the questions it forces us to ask.

Conclusion: The Modern Relevance of Descartes' Scepticism.

Descartes' philosophy, with its relentless questioning of reality, remains deeply relevant in the age of the simulation hypothesis. His scepticism, his search for certainty, and his focus on consciousness all resonate with the challenges posed by the idea that we might be living in a simulated universe. By comparing Descartes' evil demon with the Operators, we see that the fundamental questions of philosophy—what is real, what can we know, and what does it mean to exist—are as pressing today as they were in the 17th century.

The simulation hypothesis may not provide definitive answers, but it encourages us to continue Descartes' quest, to question, to doubt, and to seek the truth, even in the face of uncertainty. Whether we are living in a simulation or not, the journey of philosophical inquiry is one that shapes our understanding of ourselves and our place in the universe.

Ethical Dimensions of Simulated Existence: Free Will, Purpose, and Moral Responsibility.

The simulation hypothesis not only challenges our understanding of reality but also forces us to confront deep ethical questions about our existence and the nature of free will. If we are indeed living in a simulated universe, orchestrated and controlled by the Operators, what does this mean for our choices, our purpose, and our moral responsibilities?

Free Will in a Simulated Universe: Are We Truly Free?

One of the most profound ethical dilemmas posed by the simulation hypothesis is the question of free will. If our reality is programmed, if our actions are determined by a complex set of algorithms created by the Operators, then can we genuinely claim to have free will? Or are we simply following a predetermined path, our choices and actions mere illusions of autonomy?

From a philosophical standpoint, free will is the capacity to choose between different possible courses of action unimpeded. It is a cornerstone of many ethical systems, underpinning concepts of moral responsibility, justice, and individual agency. In a simulated reality, however, this concept becomes murky. If the Operators can manipulate or even predict our actions, then are we truly making choices, or are we like characters in a video game, following a script we cannot see?

Consider a scenario where every decision you make—whether to turn left or right, to say yes or no—is part of a pre-written program. You might feel that you are making choices, but those choices have already been encoded into the simulation's structure. This deterministic view of existence can be unsettling, suggesting that all our struggles, our achievements, and even our moments of self-realization are preordained.

Yet, even within this deterministic framework, there may be room for free will, or at least the perception of it. Perhaps the Operators have designed the simulation to be open-ended, allowing us to make choices within a set of parameters. This would mean that while the broad strokes of our existence are predetermined, we still have the freedom to navigate the details. In this sense, free will might be a kind of sandbox within the larger structure of the simulation—a space where we can exercise creativity and agency, even if the ultimate outcome is beyond our control.

The Purpose of the Simulation: Experiment, Entertainment, or Enlightenment?

If we are part of a simulation, then why does it exist? What is the purpose behind the elaborate construction of a universe, complete with conscious beings capable of love, pain, joy, and sorrow? The answer to this question has profound ethical implications, as it speaks to the intentions of the Operators and the value of our simulated lives.

One possibility is that the simulation is an experiment, designed to test certain hypotheses about the nature of intelligence, morality, or social structures. In this scenario, we are subjects in a grand scientific inquiry, our actions and reactions observed and recorded by the Operators. If this is true, then our struggles and triumphs serve a greater purpose, contributing to the Operators' understanding of the universe—or perhaps even to the understanding of the beings who created them.

Another possibility is that the simulation serves as a form of entertainment. Much like we create virtual worlds and narratives for our enjoyment, the Operators might be using the simulation as a complex game, an immersive experience that allows them to explore different scenarios and outcomes. This idea raises troubling ethical questions: If our lives are a form of entertainment, does that trivialize our suffering and diminish the significance of our experiences?

A more optimistic view is that the simulation is a tool for enlightenment, both for the simulated beings and for the Operators themselves. Perhaps the simulation is a way to explore the nature of consciousness, to see how beings respond to the challenges of existence and the mysteries of the universe. If this is the case, then our lives have intrinsic value, not because they are real in a physical sense, but because they are meaningful within the context of the simulation.

Moral Responsibility of the Operators: Gods or Guardians?

If the Operators have created the simulation, then what responsibilities do they have towards its inhabitants? Are they like gods, with the power to intervene and shape our destiny? Or are they more like guardians, observing from a distance and intervening only when absolutely necessary?

If the Operators have the ability to influence or change the simulation, then they could theoretically prevent suffering, cure diseases, or eliminate injustices. Yet, the existence of pain and suffering suggests that they either choose not to intervene or that they have set the simulation to run autonomously, allowing events to unfold without interference. This raises a profound ethical question: If the Operators have the power to prevent harm but choose not to, are they morally culpable for the suffering that occurs?

This question mirrors the classic problem of evil in theology: If a benevolent and all-powerful God exists, why does suffering persist? In the context of the simulation hypothesis, the Operators might have reasons for allowing suffering, such as testing the resilience of simulated beings or exploring the moral implications of free will. However, this does not absolve them of responsibility. If they have created a world with the capacity for suffering, then they bear at least some ethical responsibility for the outcomes.

Conversely, if the Operators are more like scientists conducting an experiment, their moral obligations might be different. Scientists observing animal behaviour in the wild, for example, are generally expected not to intervene, even if the animals are in distress, to avoid disrupting the natural course of events. If the Operators take this view, they might see intervention as compromising the integrity of the simulation, preventing them from understanding the true nature of their subjects.

Human Ethics in a Simulated World: Does It Matter?

For us, as potentially simulated beings, the knowledge that we might be living in a simulation can feel disorienting. Does it change how we should act, how we should treat each other, or how we should live our lives? If our reality is artificial, does that make our moral principles any less valid?

The answer to this question depends on how we define morality. If morality is based solely on the physical reality of our actions, then the simulation hypothesis might seem to undermine its foundations. After all, if nothing is truly real, then why should we care about right and wrong?

But morality is not just about physical actions; it is also about relationships, intentions, and the effects of our choices on others. Whether we are real or simulated, the experiences we have and the emotions we feel are authentic to us. Our actions, therefore, still carry moral weight. Hurting someone in a simulated reality is no less wrong than hurting them in a physical one because the suffering they experience is real to them.

In this sense, the simulation hypothesis does not negate morality but reinforces it. If we are being observed by the Operators, then our actions take on even greater significance. We are not just living for ourselves; we are part of a larger narrative, one that may be serving a purpose we cannot yet comprehend. Whether we are characters in a story or players in a game, the choices we make define who we are, both to ourselves and to the unseen audience beyond the veil of our simulated universe.

Conclusion: Navigating the Ethical Labyrinth of the Simulation Hypothesis.

The simulation hypothesis presents a complex ethical landscape, one that challenges our notions of free will, purpose, and moral responsibility. If we are living in a simulated universe, we must reconsider what it means to make choices, to suffer, and to find meaning in our lives. We must also grapple with the nature of the Operators, whose intentions and responsibilities remain a tantalizing mystery.

Ultimately, the ethical questions raised by the simulation hypothesis are not so different from those that have confronted humanity for centuries. Whether our reality is simulated or not, we still face the same fundamental challenges: How should we live? What is our purpose? And how can we find meaning in a world that often seems indifferent to our existence?

By exploring these questions through the lens of the simulation hypothesis, we gain new perspectives on old dilemmas, deepening our understanding of what it means to be human—whether real or simulated.

The Role of Imagination in Speculating About Reality.

The human capacity for imagination is one of our most remarkable traits. It allows us to think beyond the limitations of our immediate experience, to conceive of possibilities that stretch far beyond the boundaries of the known. Imagination fuels our creativity, drives scientific innovation, and helps us grapple with the deepest questions of existence. In the context of the simulation hypothesis, imagination becomes not just a tool for speculation, but a crucial means of exploring the very nature of reality itself.

Imagination as a Tool for Philosophical Inquiry.

Throughout history, imagination has been a cornerstone of philosophical thought. It has enabled philosophers to conceive of alternate realities, thought experiments, and hypothetical scenarios that challenge our understanding of the world. Consider Plato's Allegory of the Cave, where prisoners, confined to a life of shadows, mistake those shadows for reality. It is only through imagination—both on the part of the philosopher who conceives the allegory and the prisoner who dares to look beyond the shadows—that the true nature of their existence is revealed.

The simulation hypothesis is, in many ways, a modern extension of this allegory. It invites us to imagine that our perceived reality is just a shadow on the wall, a construct created by unseen Operators. To consider this possibility requires a leap of imagination, a willingness to entertain ideas that defy our everyday experiences and challenge our deepest assumptions about the nature of existence.

Imagination, then, is not just a fanciful diversion. It is a vital tool for philosophical inquiry, enabling us to ask "what if?" and to explore the implications of those questions. It allows us to step outside the confines of our perceived reality and consider the possibility of alternate realities, whether they are simulated worlds, parallel universes, or entirely different dimensions of existence.

Speculating Beyond the Observable Universe.

The limits of scientific observation constrain our understanding of the universe. We can only see as far as our telescopes allow, measure as much as our instruments can detect, and comprehend as much as our theories can explain. But the universe is vast, and our knowledge of it is still in its infancy. Imagination, therefore, becomes a bridge between what we know and what we have yet to discover.

When we speculate about the nature of reality—whether through the simulation hypothesis, the multiverse theory, or the idea of parallel dimensions—we are engaging in a kind of imaginative exploration. We are using our minds to venture into the unknown, to consider possibilities that lie beyond the reach of our current scientific understanding. This speculative imagination is not unscientific; rather, it is a necessary precursor to scientific discovery.

Many of the greatest scientific breakthroughs began as imaginative speculations. Einstein's theory of relativity, for example, was born out of thought experiments about riding on a beam of light. Quantum mechanics, with its strange and counterintuitive principles, emerged from a willingness to imagine a world that defies classical logic. In the same way, the simulation hypothesis encourages us to imagine that our universe is not what it seems, to consider the possibility that we are part of a vast, intricate design beyond our comprehension.

Creative Storytelling as a Way to Explore the Hypothesis.

One of the most powerful ways to explore the implications of the simulation hypothesis is through storytelling. Fiction, whether in the form of literature, film, or art, allows us to delve into the possibilities of simulated realities in a way that is both accessible and profound. Stories like *The Matrix*, *Inception*, and *The Truman Show* capture the imagination because they explore the disorienting and often unsettling experience of discovering that reality is not what it seems.

But before these films brought the simulation concept to the mainstream, there was *Tron*. Released in 1982, *Tron* was groundbreaking in its portrayal of a digital world where programs are not just lines of code but sentient beings with their own society, culture, and struggles. The protagonist, Kevin Flynn, is transported into this virtual world, becoming a "User" who interacts directly with the programs. This film not only explored the idea of a simulated reality but also introduced the notion of users—analogous to the Operators in the simulation hypothesis—who exist on a higher level of reality and have the power to shape the digital world.

Tron serves as a precursor to many of the themes explored in later films like *The Matrix*. It questioned what it means to exist within a created reality, how much autonomy the inhabitants of such a world truly have, and the ethical implications of a creator stepping into their own creation. The film's depiction of the relationship between Users and Programs mirrors the dynamics we explore in this book between the Operators and the simulated beings they observe.

In *The Matrix*, the world is a meticulously crafted illusion designed to control and pacify its inhabitants, while in *Inception*, the boundaries between dreams and reality blur, challenging the protagonists' perceptions of what is real. *The Truman Show* takes a different approach, depicting a man's life as a staged television show, with every detail of his reality controlled by unseen producers. In each of these narratives, characters grapple with the revelation that their world is not what it appears to be, and they struggle to assert their autonomy in the face of overwhelming control.

Similarly, in *Tron*, the protagonist must navigate a reality that is both familiar and alien, governed by rules he doesn't fully understand and populated by entities that view him as a god-like figure. His journey to understand and ultimately escape the digital world mirrors the existential quest of those who suspect they might be living in a simulation. It asks us to consider the ethical responsibilities of creators towards their creations, a theme that resonates strongly with the philosophical questions we've been exploring.

Including *Tron* alongside these more recent films highlights the evolution of our cultural fascination with the idea of simulated realities. It shows that long before the simulation hypothesis was a topic of academic debate, storytellers were already grappling with these themes, using fiction to explore the boundaries between creator and creation, reality and illusion.

By engaging with these stories, we are able to explore the emotional, psychological, and ethical dimensions of the simulation hypothesis in a way that pure philosophical or scientific discourse cannot. These narratives help us to personalize the abstract concepts, imagining what it would be like to awaken to the realization that our world is a simulation, to struggle with the implications of that knowledge, and to find a way to live meaningfully in spite of it. In this sense, storytelling becomes a powerful tool for exploring not just the intellectual, but the human, dimensions of the simulation hypothesis.

Imagination and the Limits of Knowledge

The limits of human knowledge are vast. Despite our best efforts, there is much about the universe that we do not and perhaps cannot know. Imagination allows us to explore these limits, to push beyond the boundaries of what is currently understood, and to envision possibilities that transcend our current limitations.

The simulation hypothesis itself is a testament to the power of imagination. It challenges us to rethink our most basic assumptions about reality, to entertain the possibility that everything we know could be an illusion. This requires a willingness to step into the unknown, to question what we have always taken for granted, and to imagine a universe that is fundamentally different from the one we perceive.

But imagination must be tempered with humility. While it allows us to speculate about the nature of reality, it also reminds us that there are things we cannot know for certain. The simulation hypothesis, like any other speculative theory, is an invitation to explore the unknown, not a definitive answer to the mysteries of existence. It encourages us to remain open-minded, to question, and to wonder, even as we acknowledge the limits of our understanding.

Imagination as a Bridge Between Science and Philosophy.

The simulation hypothesis occupies a unique space at the intersection of science and philosophy. It invites us to think deeply about the nature of reality while also proposing potential avenues for scientific investigation. While proving or disproving the hypothesis may seem like an insurmountable challenge, some scientists and theorists have speculated about ways in which it might, in principle, be testable.

Testing the Hypothesis: Searching for Anomalies in the Universe.

One proposed method for testing the simulation hypothesis involves searching for anomalies or inconsistencies in the physical laws of the universe that might indicate we are living in a constructed reality. For example, just as a video game might have rendering errors or glitches, our universe might exhibit certain "signs" that it is being simulated. These could include unexpected deviations in the behaviour of particles, limits to the resolution of space and time, or unexplainable patterns in cosmic radiation. If such anomalies were found, they could provide indirect evidence that our universe has underlying computational limitations, similar to those found in a simulated environment.

Computational Constraints and the "Resolution" of the Universe.

Another approach involves looking for evidence of computational constraints in the fabric of reality. In a digital simulation, the universe would have a finite resolution, much like the pixels on a screen. Physicists have suggested that we could, in theory, detect this finite resolution in the fabric of space-time, perhaps through the limitations in the energy or information that can be stored in a given region of space. This idea is inspired by the Planck scale, which sets fundamental limits on the smallest measurable lengths and times in the universe. If it were possible to find a minimum "pixel size" in the structure of space-time, it might point to the universe being rendered in discrete units, much like a digital simulation.

The Holographic Principle and Information Paradox.

The holographic principle, which suggests that the information content of all the matter in a given volume of space can be represented as encoded on the boundary of that space, offers another potential avenue for testing the simulation hypothesis. If the universe is fundamentally a two-dimensional information structure projected as a three-dimensional reality, this could imply that our reality operates like a highly sophisticated computational process. Investigating the nature of this information and how it is preserved or transformed, particularly in extreme conditions like near black holes, might provide insights into whether we are living in a simulated environment.

Quantum Physics and Digital Reality.

Quantum mechanics, with its probabilistic nature and the phenomenon of entanglement, has also been suggested as a possible area for testing the simulation hypothesis. Some researchers have speculated that the "weirdness" of quantum physics could be a reflection of underlying computational processes. For example, the way particles exist in superpositions or the "spooky action at a distance" seen in entanglement might be more easily explained if our universe were a simulation. The idea here is that the fundamental randomness and apparent "glitches" at the quantum level could be signs that reality is being rendered or computed on demand.

Theoretical Limits and Practical Challenges.

While these ideas are fascinating, it's important to acknowledge the immense practical challenges involved. Even if we were to detect such anomalies, proving that they result from a simulated reality, rather than being natural features of a non-simulated universe, would be extraordinarily difficult. Nonetheless, these speculative avenues highlight how imagination can inspire scientific inquiry, bridging the gap between philosophical speculation and empirical investigation.

Ultimately, whether or not the simulation hypothesis is ever testable in a definitive way, its value lies in the questions it raises and the perspectives it offers. It encourages us to think beyond the familiar and to consider possibilities that stretch the boundaries of both science and philosophy. In this way, imagination becomes a bridge, linking our search for knowledge with our deepest philosophical inquiries.

Conclusion: The Power and Purpose of Imagination.

Imagination is a uniquely human faculty, one that allows us to explore the unknown, to question the nature of reality, and to speculate about the deepest mysteries of existence. In the context of the simulation hypothesis, imagination is not just a tool for speculation; it is a means of navigating the profound ethical, philosophical, and existential questions raised by the possibility that our universe is a simulation.

Whether we are imagining alternate realities, crafting stories that explore the boundaries between illusion and reality, or devising experiments to test the limits of our knowledge, imagination is at the heart of our inquiry. It enables us to ask "what if?" and to explore the implications of those questions in ways that are both intellectually rigorous and deeply human.

The simulation hypothesis, with all its unsettling and intriguing possibilities, invites us to use our imagination to its fullest. It challenges us to think beyond the confines of our current understanding, to question the nature of reality, and to seek answers to questions that may never be fully resolved. In doing so, it reminds us of the power and purpose of imagination—to explore, to wonder, and to seek meaning in a universe that is, in the end, as mysterious and limitless as our capacity to imagine it.

End of chapter 8.

Chapter 9: Alternate Dimensions: Fiction, Physics, and the Simulation Hypothesis

Introduction: The Allure of Alternate Dimensions

Alternate dimensions, parallel worlds, and multiverses have become essential themes in both science fiction and theoretical physics, captivating audiences and thinkers alike. From the mind-bending realities of *Doctor Strange* and *Into the Spider-Verse* to the dark, alternate histories of *The Man in the High Castle* and the chaotic, dimension-hopping antics of *Rick and Morty*, these narratives push the boundaries of our understanding of reality. They challenge us to consider: What if our universe is just one of many, each with its own unique rules and histories?

This chapter explores how these portrayals reflect deeper philosophical and scientific ideas, integrating them with the simulation hypothesis. Could it be that the Operators are not just managing a single simulation, but an entire multiverse of alternate dimensions? And how do the theories of physics support the notion of multiple realities, each interconnected and potentially accessible? By weaving together fictional narratives and scientific speculation, we will delve into the potential existence of alternate dimensions and their implications for our understanding of reality.

Alternate Dimensions in Popular Culture: Expanding the Narrative Universe

The concept of alternate dimensions has been explored extensively in popular culture, particularly in franchises like Marvel and DC, but also in other series that tackle the multiverse in unique ways. Each of these narratives offers a different take on what alternate dimensions could look like and how they might interact with our own.

In the *Marvel Universe*, films like *Doctor Strange* and *Into the Spider-Verse* depict alternate dimensions as vibrant, diverse worlds, each governed by its own set of rules. *Doctor Strange* navigates through realities where time loops endlessly, gravity operates in strange ways, and the very fabric of space bends and folds. These realities, though fantastical, reflect a deep-seated human fascination with the possibility that our universe is not alone. *Into the Spider-Verse* takes this a step further by bringing together multiple versions of Spider-Man, each from a different dimension. The film suggests that every possible version of a person could exist somewhere, each living out their own unique story.

Similarly, *Rick and Morty* presents a chaotic, often humorous take on the multiverse, where Rick, the brilliant but nihilistic scientist, drags his grandson Morty through countless alternate realities. Each episode explores bizarre worlds with their own internal logic, from a dimension populated by sentient furniture to one where everyone speaks in gibberish. The show uses these absurd scenarios to satirize the idea of infinite possibilities and to question the significance of individual choices and events. If there are infinite versions of yourself making every possible choice, does any single decision truly matter?

The Man in the High Castle offers a more grounded, yet equally compelling portrayal of alternate dimensions. It imagines a world where the Axis powers won World War II, drastically altering the course of history. The series introduces the concept of alternate realities through mysterious films that show different versions of events, suggesting that every choice creates a divergent timeline. This idea resonates with the multiverse theory, where every possible outcome of every decision exists in its own separate reality.

Stranger Things takes a different approach by introducing the "Upside Down," a dark, parallel dimension that mirrors our own but is twisted and corrupted. The show explores the consequences of opening a portal between dimensions, suggesting that even brief interactions between realities can have catastrophic consequences. The Upside Down is both fascinating and terrifying, a place that feels familiar yet profoundly alien, embodying the fear of the unknown that underlies many explorations of alternate dimensions.

These narratives do more than entertain. They challenge us to think about the nature of reality and the possibilities that lie beyond our perception. They suggest that our universe could be one of many, each with its own unique properties, histories, and inhabitants. But how do these fictional depictions align with the scientific theories of alternate dimensions?

Theoretical Physics and Alternate Dimensions: From String Theory to the Multiverse

The idea of alternate dimensions is not just a whimsical notion of fiction, but a serious subject rooted in theoretical physics. For decades, scientists have explored the possibility that our universe is just one within an expansive multiverse, a collection of countless universes that coexist alongside our own. These universes, governed by their own sets of physical laws and constants, may be hidden from our perception yet are intrinsic to the fabric of reality.

At the forefront of these explorations is string theory, which has revolutionised our understanding of the cosmos. String theory posits that the fundamental building blocks of the universe are not particles, as previously believed, but incredibly small, vibrating strings of energy. These strings do not merely oscillate in the familiar dimensions of space and time, but in additional, hidden dimensions that remain imperceptible to us.

String theory is often considered one of the most promising candidates for a "theory of everything," offering a framework that could unify all the fundamental forces of nature—gravity, electromagnetism, and the strong and weak nuclear forces—into one coherent theory. At its core, string theory opens up the possibility of extra dimensions beyond the four-dimensional spacetime we experience. These higher dimensions, though hidden from view, play a crucial role in determining the properties of the strings that define our universe's fundamental particles.

This chapter will trace the evolution of string theory from its inception as a bold hypothesis in the late 20th century, through the profound developments that led to the emergence of M-theory, a unifying framework that suggests all versions of string theory are different expressions of the same underlying reality. As theoretical physicists probed deeper into the implications of string theory, they proposed the existence of eleven dimensions, not just the four we are accustomed to. These additional dimensions, though compactified or "curled up" at minuscule scales, are essential for the consistency and mathematical coherence of the theory.

Perhaps the most mind-bending implication of string theory is the notion that these extra dimensions might contain entire other universes. Our universe could be a three-dimensional "brane" (short for membrane) floating in a higher-dimensional space, often referred to as the "bulk." In this framework, other branes might exist parallel to ours, each corresponding to a different universe with its own set of physical laws. These universes could be separated by vast expanses of higher-dimensional space, or they might be as close as a fraction of a millimetre away, yet undetectable due to the constraints of our three-dimensional perception.

Such ideas also bring us to brane cosmology, a model within string theory that portrays our universe as one of many branes. Brane cosmology, championed by physicists like Lisa Randall and Raman Sundrum, offers intriguing explanations for some of the most puzzling aspects of our reality, such as why gravity is so much weaker than the other forces. According to this model, gravity might be able to "leak" into the higher-dimensional bulk, while the other forces are confined to the brane we inhabit, making gravity appear weaker in comparison.

Theoretical physics does not stop with string theory and brane cosmology when it comes to alternate dimensions. Concepts like the multiverse hypothesis suggest that every possible outcome of every event spawns a new universe, leading to an infinite array of realities. These alternate dimensions, far from being mere science fiction, are grounded in serious mathematical exploration. They challenge our understanding of space, time, and reality, pushing the boundaries of what might be possible.

In addition to expanding our perception of the universe, these theories have profound implications for philosophical questions and even the simulation hypothesis. If there are multiple dimensions and universes, could they be part of a grander simulation? Could the Operators, the creators of this reality, be managing not just one universe, but an entire multiverse, each with its own unique set of rules and interactions? These questions invite us to reconsider the very nature of existence and how physics and speculative thought intersect.

String Theory and Higher Dimensions

String theory, one of the leading candidates for a theory of everything, proposes that the fundamental particles of our universe are not point-like objects but tiny, vibrating strings. These strings can vibrate in multiple dimensions, beyond the familiar three dimensions of space and one of time. The theory has been developed and championed by some of the most prominent theoretical physicists of our time, each contributing to its evolution and the exploration of its implications.

The Birth of String Theory: Pioneers and Early Development.

String theory emerged in the late 1960s as an attempt to describe the strong nuclear force, but it soon evolved into a more comprehensive framework that could potentially unify all the fundamental forces of nature. The early development of string theory is attributed to physicists such as **Gabriele Veneziano**, who discovered the mathematical framework that led to the initial formulation of the theory, and **Leonard Susskind**, who was among the first to recognize its potential to describe particles as one-dimensional strings.

In the 1980s, the theory underwent a significant transformation with the introduction of superstring theory, which incorporates supersymmetry—a proposed symmetry between bosons and fermions. This period, known as the first superstring revolution, was driven by the work of physicists like **Michael Green** and **John Schwarz**, whose research showed that superstring theory could be consistent and free of certain anomalies, making it a promising candidate for unifying the forces of nature.

The Eleven Dimensions: The Second Superstring Revolution.

The second superstring revolution, in the mid-1990s, introduced the concept of M-theory, which suggests that the five different versions of string theory are actually different manifestations of a single, underlying theory. This unification was largely driven by the work of **Edward Witten**, a theoretical physicist who is often considered one of the leading figures in the field. Witten's work revealed that M-theory requires not just ten dimensions, as originally proposed by string theory, but eleven dimensions—ten of space and one of time.

In this expanded framework, the additional dimensions are typically "curled up" or compactified at scales so small that they are imperceptible to us. However, these hidden dimensions are essential for the mathematical consistency of the theory. They could, in theory, host entire universes with their own distinct physical laws, vastly different from the four-dimensional space-time we experience.

Brane Cosmology: Our Universe as a Membrane

One of the key concepts that emerged from string theory is the idea of brane cosmology, where our universe is seen as a "brane" (short for membrane) floating in a higher-dimensional space known as the "bulk." This concept was further developed by physicists such as **Lisa Randall** and **Raman Sundrum**, who proposed the Randall-Sundrum models. These models describe how our universe could be a three-dimensional brane embedded in a higher-dimensional space, potentially explaining why gravity is so much weaker than the other fundamental forces.

According to this view, other branes could exist in the bulk, each representing a different universe with its own distinct properties. These branes might occasionally interact, or they could be entirely isolated, their presence detectable only through gravitational effects or other subtle influences on the structure of space-time. This idea not only supports the existence of alternate dimensions but suggests that they are intimately connected to the very fabric of our own reality.

Challenges and Controversies.

While string theory and its extensions, such as M-theory, offer a tantalizing glimpse into a universe with many more dimensions than we perceive, they are not without controversy. One of the main criticisms is that these theories have yet to produce testable predictions that could be verified through experiments. This has led some physicists, like **Lee Smolin** and **Peter Woit**, to question whether string theory can ever be considered a true scientific theory in the Popperian sense, as it currently lacks empirical validation.

Despite these challenges, string theory remains a dominant framework in theoretical physics, providing a rich mathematical structure that continues to inspire new ideas and potential connections between the known and unknown aspects of our universe. It is a testament to the power of imagination and mathematical beauty, pushing the boundaries of our understanding and suggesting that our reality might be far more complex and interconnected than we ever imagined.

Implications for the Simulation Hypothesis.

In the context of the simulation hypothesis, string theory and its higher-dimensional framework offer a fascinating perspective. If our universe is indeed a brane floating in a higher-dimensional bulk, then the Operators could be using these extra dimensions as part of their simulation architecture. Each dimension could represent a different simulation, a different set of parameters and rules designed to explore various aspects of reality. The interplay between branes could be a way for the Operators to observe how different universes interact, to test the robustness of their simulations, or to explore the consequences of cross-dimensional influences.

This perspective bridges the gap between the speculative nature of the simulation hypothesis and the rigorous mathematical framework of string theory, suggesting that the hidden dimensions and parallel universes proposed by physicists could be the underlying structure of the multiverse managed by the Operators. Whether these dimensions are physical realities or digital constructs, they challenge us to rethink our understanding of space, time, and the very nature of existence.

The Multiverse Hypothesis.

The multiverse hypothesis expands on this concept by proposing that there are countless universes, each with its own set of physical constants and laws of nature. These universes could be the result of quantum fluctuations, with each new universe branching off from another like leaves on a cosmic tree. In this view, every possible outcome of every event creates a new universe, leading to an infinite number of realities, each slightly different from the last.

The Many-Worlds Interpretation of quantum mechanics provides a foundation for this idea. It suggests that every time a quantum event occurs, the universe splits into multiple branches, each representing a different outcome. In one universe, Schrödinger's cat is alive; in another, it is dead. This interpretation of quantum mechanics implies that all possible histories and futures are real, each existing in its own parallel universe.

Wormholes and Dimensional Gateways.

If alternate dimensions exist, how could they interact with our own? One possibility is through wormholes, hypothetical bridges that connect different points in space-time. A wormhole could, in theory, link not just two distant locations in our universe, but two entirely different universes or dimensions. While purely speculative, the concept of wormholes offers a tantalizing glimpse of how beings or information might travel between dimensions, much like the portals seen in *Doctor Strange* or *Stranger Things*.

In the context of the simulation hypothesis, wormholes and other dimensional gateways could be seen as backdoors built into the simulation, allowing the Operators to move between different simulations or to observe how different realities interact. These hypothetical bridges between dimensions could be part of the Operators' design, a way to explore the consequences of inter-dimensional interactions and to test the resilience and adaptability of the simulated beings.

The Operators and the Multiverse: Creating and Managing Alternate Realities.

If the Operators are the architects of our reality, as suggested by my take on the simulation hypothesis, then their role might extend beyond creating and managing a single universe. They could be responsible for an entire multiverse of simulations, each designed to explore different scenarios, test different hypotheses, or simply to observe the infinite variations of existence.

The Purpose of a Simulated Multiverse.

Why would the Operators create multiple simulations? One possibility is that they are conducting a series of experiments, each designed to test different variables. In one dimension, they might alter the laws of physics to see how intelligent life adapts. In another, they might change the fundamental nature of consciousness or morality to explore different philosophical questions. Each dimension would represent a different experiment, a different way of exploring the potentialities of existence.

Alternatively, the Operators might be storytellers, crafting an epic narrative that spans multiple dimensions. Just as Marvel and DC use the multiverse to tell stories that explore different versions of their characters and worlds, the Operators might use their simulated multiverse to explore different versions of humanity, different outcomes of history, and different possibilities for the future. Each dimension would be a new chapter in the story, a new layer in the intricate tapestry of reality.

Cross-Dimensional Interactions and Bleed-Through.

The concept of "bleed-through" is a common trope in fiction, where elements from one dimension affect another. In the context of the simulation hypothesis, this could represent unintended interactions between different simulations. Perhaps a character or event from one dimension crosses over into another, creating a ripple effect that alters the course of both realities. These interactions could be accidental, the result of a glitch in the simulation, or intentional, designed by the Operators to observe the consequences of such interactions.

This idea also raises the possibility of civilizations within the simulation discovering these dimensional gateways and attempting to travel between them. If a civilization within the simulation becomes advanced enough to understand its own nature, it might try to hack the simulation, to break through the barriers between dimensions and explore the multiverse created by the Operators. Such an event would be a profound challenge to the simulation's integrity, a cosmic jailbreak that could have far-reaching consequences.

Philosophical and Existential Implications: Living in a Multiverse

The existence of alternate dimensions, whether fictional or theoretical, challenges our understanding of identity, morality, and the nature of existence. If there are infinite versions of ourselves, living out every possible variation of our lives, what does that mean for our sense of self and our sense of purpose?

The Nature of Identity in a Multiverse

If every possible version of you exists somewhere in the multiverse, which one is the "real" you? Are you the sum of all your possible selves, or is each version a distinct individual, living out their own unique story? This question challenges our traditional notions of identity and individuality. It suggests that we are not singular beings, but part of a vast, interconnected web of possibilities.

This idea is explored in *Rick and Morty*, where different versions of Rick and Morty interact with each other, sometimes with devastating consequences. The show satirizes the idea that if there are infinite versions of yourself, then no single version truly matters. This nihilistic perspective is contrasted with the emotional struggles of the characters, who grapple with the existential implications of their own expendability.

Moral Responsibility Across Dimensions.

If the Operators are creating and managing multiple dimensions, do they have a moral responsibility to the inhabitants of each one? Are they obligated to prevent suffering, or are they free to experiment as they please, knowing that each dimension is just one of many?

For the inhabitants of the simulation, this raises profound ethical questions. If you knew that every possible version of yourself existed somewhere, would that change how you lived your life? Would you be more willing to take risks, knowing that even if you failed, there would be a version of you that succeeded? Or would you feel overwhelmed by the weight of infinite possibilities, paralyzed by the knowledge that every choice creates a new branch in the tree of your existence?

Finding Meaning in a Simulated Multiverse.

In a universe where everything is possible, how do we find meaning? The simulation hypothesis, combined with the idea of a multiverse, suggests that our reality is just one of many, each as real or as unreal as the next. This can be a disorienting thought, but it also offers a new way of thinking about our place in the cosmos.

Perhaps the meaning of our existence is not found in any single reality, but in the sum of all realities. Each version of ourselves, each possible world, contributes to the greater whole, to the grand experiment or story being crafted by the Operators. In this view, we are not isolated beings, but part of a vast, interconnected network of realities, each playing its part in the unfolding drama of existence.

Conclusion: Navigating the Multiverse of Simulations

The exploration of alternate dimensions, both in fiction and in theoretical physics, offers a rich framework for understanding the simulation hypothesis. Whether through the fantastical journeys of *Doctor Strange*, the dark alternate histories of *The Man in the High Castle*, or the chaotic multiverse of *Rick and Morty*, these stories challenge us to think beyond the limits of our own reality and to consider the infinite possibilities that lie beyond.

In the context of the simulation hypothesis, these alternate dimensions are not just speculative ideas, but potential realities crafted by the Operators. They represent different ways of exploring the nature of existence, testing the limits of intelligence, and understanding the infinite possibilities of the universe. By integrating the imaginative explorations of popular culture with the rigorous speculations of theoretical physics, we gain a deeper understanding of what it might mean to live in a simulated multiverse, where each dimension is a new chapter in the story of existence.

End of chapter 9.

Chapter 10: A Universe Designed by Code

Reality as a Program – Awaiting an Upgrade

The universe, as we understand it, appears to be governed by immutable laws—rules that have remained constant throughout the history of existence. But what if these laws, the very fabric of reality, are not as fixed as they seem? Imagine a scenario where the universe is a vast, sophisticated program, and the constants of nature are parameters set by the Operators to create a specific version of reality. Now, envision that this program is not static, but subject to updates, much like the operating systems or games we use every day. What would it mean for us, for civilizations, and for the cosmos itself, if the Operators decided it was time for an upgrade?

In this chapter, we explore the possibility that the universe could be fundamentally altered by its creators, who might choose to modify the simulation to explore new avenues of existence or to test the adaptability of its inhabitants. Such an upgrade would be akin to releasing a new version of reality—one where the rules are different, the possibilities expanded, and the stakes even higher.

Programmable Parameters: A Future Upgrade?

We've touched on the idea that the universe operates according to a set of parameters—fundamental constants that shape everything from the structure of atoms to the expansion of galaxies. These parameters, like the speed of light, the gravitational constant, and Planck's constant, are the foundation upon which all of physics is built. They appear unchanging and universal, but what if they are more like settings in a vast, cosmic program, adjustable at the will of the Operators?

Imaginable Modifications: Preparing for a New Reality.

Imagine waking up one day to find that the speed of light has doubled. The very nature of space-time would change, affecting everything from the way we perceive distance to the stability of matter itself. Entire constellations might shift, and the boundaries of what is possible could be pushed far beyond our current understanding. Such a modification would not be random; it would be a deliberate act by the Operators, a reconfiguration of the simulation to facilitate a new phase of their experiment.

Why Would the Operators Choose to Update?

If the universe is an ongoing experiment, there might come a time when the Operators feel that the current version has run its course. Perhaps they wish to explore how advanced civilizations cope with drastically different physical laws. Or maybe they seek to unlock new forms of existence and consciousness that the current configuration cannot support. Much like a software developer pushing an update to fix bugs or introduce new features, the Operators might upgrade the universe to refine or enhance the simulation.

Potential Changes and Their Impact

- **Speed of Light ©:** If the speed of light were increased, communication and travel would become far more feasible over interstellar distances, potentially leading to a new era of galactic civilizations. Conversely, a reduction could isolate civilizations further, creating a more fragmented, introspective universe.
- **Gravitational Constant (G):** Altering the strength of gravity could lead to the formation of entirely new types of stars and galaxies. A stronger gravitational pull might allow for more compact and dense structures, while a weaker force could create sprawling, diffuse formations, altering the very structure of the cosmos.

- **Planck's Constant (h):** Changing the scale at which quantum effects are significant could lead to a universe where quantum phenomena are visible on a macroscopic scale, blurring the line between the quantum and classical worlds. This could give rise to entirely new forms of technology and consciousness.

An Ever-Evolving Simulation.

Such changes would be far more significant than any natural catastrophe or cosmic event we have ever imagined. They would represent a shift not just in the content of the universe, but in its very framework. Civilizations, if they were to survive, would need to adapt rapidly to the new rules of reality. They might develop technologies we cannot currently fathom, or perhaps even evolve into forms of life suited to the new conditions.

The Operators, observing from beyond the simulation, would gain valuable insights into the nature of adaptability and survival. Each upgrade would test the limits of creativity, resilience, and intelligence, pushing the inhabitants of the simulation to new heights—or eliminating those unable to cope with the new version of reality.

Can We Decipher the Code of the Simulation?

Imagine a future where civilizations have advanced to the point where they begin to grasp the very fabric of the universe. They're no longer just studying stars and galaxies; they're delving into the hidden logic that governs all of existence. This isn't merely a scientific quest; it's akin to trying to reverse-engineer a vast, cosmic software. Each discovery they make, whether it's about quantum entanglement or the nature of dark matter, is like uncovering a snippet of the source code that underpins reality itself.

For such civilizations, the stakes couldn't be higher. If they can decipher even a fraction of this cosmic code, they might gain the ability to anticipate the Operators' next move. After all, just as we prepare for software updates, they could prepare for the universe's version of an upgrade. They might even develop technologies capable of detecting subtle shifts in the simulation, such as minor fluctuations in the speed of light or anomalies in the structure of space-time. It would be as if they were listening for the telltale hum of the universe's machinery, trying to catch the Operators in the act of preparing a new version of reality.

But what if they go further? What if they build their own version of a universal Turing machine—a hypothetical device capable of simulating any possible universe? With such a machine, they could explore countless hypothetical scenarios, experimenting with different versions of the fundamental constants and observing the outcomes. Imagine a civilization running countless simulations of their own reality, trying to predict what might happen if, say, the gravitational constant were halved or if the speed of light were suddenly doubled. They would be preparing for a cosmic upgrade in the most literal sense, using their simulations as training grounds for a future where the universe itself is about to change.

Of course, this would be a double-edged sword. Even if they could predict an upgrade, they would still be at the mercy of the Operators' true intentions, which remain a mystery. The Operators might introduce changes that no amount of preparation could foresee—entirely new dimensions, forces that defy our understanding, or even a complete reset of the universe's history. For all their technological prowess, these civilizations would still be fumbling in the dark, hoping that their preparations would be enough to survive whatever comes next.

In this light, the quest to understand the universe becomes more than a scientific endeavour. It's a race against time, a desperate attempt to gain mastery over their destiny before the very nature of reality is altered beyond recognition. The risk is immense: to succeed would be to unlock a new level of existence, to fail would mean extinction, erased by the Operators as just another failed experiment.

Exploring Quantum Computing and Space-Time Manipulation

Quantum computing is perhaps the closest we've come to building machines that think like the universe itself. While classical computers, like the ones we use today, struggle with simulating even a handful of particles at the quantum level, quantum computers excel at this task. They operate on the same principles that govern the very fabric of reality, making them the perfect tools for civilizations striving to understand and anticipate the Operators' next move.

Picture a civilization harnessing the power of quantum computers not just to simulate the behaviour of particles, but to experiment with different versions of the universe's code. With these machines, they could explore how subtle tweaks to the fundamental constants might ripple through the fabric of space-time, changing the rules of the game in ways we can barely imagine. They could run simulations that mimic the Operators' potential upgrades, testing their own adaptability to a universe where the speed of light is no longer a constant, or where gravity behaves in ways that defy current understanding.

But even this is just the beginning. Imagine if these civilizations mastered the art of space-time manipulation. The next version of the simulation might introduce changes that warp the very nature of space and time, making the familiar laws of physics seem like relics of a simpler era. In such a universe, the distance between two points might not be fixed, or time might flow differently in different regions. How would a civilization navigate such a reality?

They might develop spacecraft capable of adapting to these fluctuations, using quantum algorithms to calculate the most efficient routes through a constantly shifting landscape. Or they might build habitats that exist outside of conventional space-time, safe havens where they could ride out the storm of an upgraded universe. For these civilizations, surviving the next version of the simulation would require more than just advanced technology; it would demand a fundamental rethinking of what it means to exist in a reality where the rules are in constant flux.

And what if the Operators introduced entirely new dimensions, opening up pathways to realms beyond our current comprehension? Such an upgrade would be like discovering that the world isn't just three-dimensional, but that there are hidden layers, unseen and unexplored, that have always been there, just beyond our perception. For a civilization that's prepared, these new dimensions would represent a vast frontier, full of untapped potential and unimaginable risks.

Yet, for all their preparation, these civilizations would still be venturing into the unknown. The next version of the simulation could introduce phenomena that defy all previous experience, new forms of matter and energy that change the nature of existence itself. In such a universe, the only constant would be change, and the ability to adapt would become the most valuable asset of all.

The challenge, then, is not just to survive, but to thrive in a reality where the very fabric of existence is malleable. It's a test of creativity, resilience, and intelligence, as the inhabitants of the simulation struggle to understand and navigate a universe that is, in every sense, beyond their control. And as they push the boundaries of what is possible, they bring themselves one step closer to understanding the Operators' true purpose, even if they may never fully grasp it.

The Future of Physics: Beyond the Simulation.

As we stand on the precipice of understanding, one thing becomes increasingly clear: the universe we know might only be the first chapter in a much larger story. The simulation, with all its intricacies and constraints, could be merely a starting point—a controlled environment designed to nurture and challenge its inhabitants. But what if this version of reality is destined to be replaced, much like an outdated operating system? What lies beyond could be a future that defies all current understanding, where the rules of existence are rewritten in ways that are as thrilling as they are terrifying.

Imagine a universe where the constants of nature are no longer constant, where the speed of light, the strength of gravity, and even the flow of time are fluid, shifting according to the whims of a new and more advanced version of the simulation. In this reality, the very fabric of space-time could be woven into patterns we cannot currently conceive. Planets might orbit stars in dimensions beyond our comprehension, while entire galaxies could fold and unfold like origami in a multidimensional space.

Such a reality would be a playground for advanced civilizations, a cosmic sandbox where the possibilities are limited only by the imagination of the Operators—or perhaps, by the creativity of those who have managed to transcend the current version of the simulation. In this upgraded universe, the tools of physics as we know them would be obsolete. Instead of rockets and spaceships, civilizations might travel instantaneously by manipulating the very essence of space-time, stepping from one corner of the cosmos to another as easily as we cross a room.

And what of life itself? Could the Operators introduce new forms of existence, entities that do not rely on matter or energy as we understand them? These beings might be made of pure information, living in the interstices of the simulation, able to traverse the layers of reality with a freedom we can only dream of. They could be the true heirs of the upgraded universe, evolving beyond the need for physical form, experiencing existence as a series of conscious states, flowing through the simulation like waves on an infinite ocean of possibilities.

For those left behind in the older version of the simulation, the upgraded universe might appear as a series of impossible phenomena, glimpsed only in the rarest of moments. A sudden flash of light as a civilization steps through a wormhole into another dimension, or a ripple in the fabric of space-time as a galactic-scale intelligence rewrites the rules of reality to suit its needs. To us, these would be miracles, inexplicable and awe-inspiring, the faint echoes of a future we can barely imagine.

But what if we, too, could ascend? What if the Operators, in their wisdom—or perhaps their curiosity—have left a path for us to follow, a series of clues hidden within the very structure of the simulation? If we can decipher these clues, we might gain access to the tools and knowledge needed to bridge the gap between versions, to become part of the upgraded universe rather than a relic of the past.

This path might lead us to confront our deepest assumptions about the nature of reality. We would need to embrace the idea that everything we know could change in an instant, that the rules we have relied on for millennia might be swept away like sandcastles in the tide. To prepare for this future, we would need to evolve—not just technologically, but spiritually and intellectually, developing a flexibility of mind and spirit that allows us to adapt to the most profound changes imaginable.

The next version of the simulation could be a test, not just of intelligence, but of adaptability and wisdom. It might challenge us to let go of everything we think we know, to trust in our ability to navigate a world where the ground shifts beneath our feet and the sky above us is an ever-changing tapestry of possibilities. It would be a world of endless potential, where the only limit is our own willingness to embrace the unknown.

And beyond this, what lies in the Operators' ultimate plan? Could the upgraded simulation be a step towards something even greater, a preparation for a reality that transcends even the most advanced version of the simulation? Perhaps this entire process, this sequence of versions and upgrades, is part of a grand experiment, designed to explore the nature of existence itself, to push the boundaries of what is possible, and to discover what lies beyond the furthest horizon of imagination.

In this vision, the Operators are not merely observers or programmers; they are explorers, seeking to understand the very essence of being. And we, as inhabitants of their creation, are part of that quest. Each upgrade, each new version of the simulation, is a chapter in a story that spans not just one universe, but countless realities, each more wondrous and complex than the last.

As we near the end of this chapter, one thing becomes clear: the journey is far from over. The Operators may choose to update the simulation, to introduce a new version that challenges us in ways we cannot yet foresee. But even if they do, the spirit of exploration, the drive to understand and transcend, will continue. We are, after all, part of the same cosmic quest, bound together by the desire to know and to be.

The future of physics, then, is not just about understanding the universe as it is, but about preparing for the universe as it might become. It is about reaching beyond the limits of our current knowledge, embracing the possibility of change, and daring to imagine a reality that is as boundless as the imagination of those who created it.

The Big Bang: Echoes of a Previous Update.

As we bring this journey through the simulation hypothesis to a close, we find ourselves contemplating one of the greatest mysteries of all: the origin of the universe itself. The Big Bang, that cataclysmic event that set everything in motion, has been the cornerstone of cosmology for nearly a century. But what if the Big Bang wasn't the beginning, but merely a beginning? What if it was the result of a previous version of the simulation being reset or updated, the moment when the Operators decided to rewrite the very fabric of reality?

In this view, the cosmic microwave background radiation—those faint whispers of energy that permeate the universe—is more than just the afterglow of a fiery birth. It's the lingering trace of a grand cosmic update, the residual energy left over from when the old version of the simulation was overwritten by the new. Like the faint hum of an operating system booting up, the cosmic radiation is the echo of that monumental reboot, a reminder that our universe, as vast and ancient as it seems, might be just the latest iteration of a much larger project.

Imagine, for a moment, the Operators gathered at the edge of the previous universe, preparing to initiate the upgrade. Perhaps they had reached the limits of what could be achieved within the old simulation. Civilizations had risen and fallen, galaxies had formed and dissolved, and the fundamental parameters of reality had been tested to their breaking point. The experiment, in its current form, had run its course. It was time for something new.

And so, with a cosmic command, the reset button was pressed. In an instant, the old universe collapsed into a singularity, an infinitesimal point of unimaginable density and potential. Then, in a flash of pure creation, the new universe exploded into being. Space and time unfurled like a vast cosmic canvas, painted with the strokes of new physical laws and constants. The speed of light, the gravitational constant, the strength of electromagnetic forces—all recalibrated, set to values that would give rise to a new reality, one with its own unique possibilities and challenges.

The Big Bang, then, was not the birth of everything, but the birth of this version of everything. It was a reboot, a fresh start for the simulation, designed to explore new frontiers and test the limits of what is possible. And we, the inhabitants of this universe, are living in the aftermath of that update, exploring the parameters and possibilities set forth by the Operators in their quest for understanding.

A Universe Poised for Another Update?

If the Big Bang was the result of a previous update, then what does that mean for us? Could another such event be on the horizon, an impending cosmic upgrade that will once again rewrite the rules of reality? The Operators, observing from beyond the confines of the simulation, might be waiting to see how far we can push the boundaries of this version of the universe. When we reach the limits of this reality, will they decide to initiate another update, to push us into a new and even more challenging version of the simulation?

The thought is both exhilarating and terrifying. Imagine a future where we have deciphered the code of the current simulation, where we have mastered the manipulation of space-time and unlocked the secrets of consciousness itself. We stand at the precipice of understanding, on the verge of breaking through the final firewall that separates us from the Operators. And then, just as we are about to take that final step, the universe begins to change.

The speed of light shifts, the constants of nature begin to fluctuate, and the very fabric of space-time twists and bends in ways we cannot comprehend. A new version of the simulation is coming online, one that will challenge everything we know, everything we are. In that moment, we would face a choice: to adapt, to evolve, to embrace the new reality and all its possibilities—or to be left behind, a relic of the old universe, erased by the march of cosmic progress.

The Silent Universe Reconsidered.

As we look out into the cosmos, we are struck by its silence. The vastness of space seems empty, devoid of the voices of other civilizations. But perhaps this silence is not a sign of absence, but of waiting. Waiting for the next great update, the next version of reality that will sweep through the universe, transforming everything in its wake.

The Fermi Paradox, that haunting question of why we have not yet encountered other advanced civilizations, might be a symptom of our place within the simulation. We are in a version of the universe that is still maturing, still exploring its potential. Other civilizations, if they exist, might be waiting for the same thing we are: the next great upgrade, the moment when the Operators once again press the cosmic reset button and a new universe bursts into being.

A Profound and Explosive Finish.

And so, we end with a vision of the universe poised on the brink of transformation. The Big Bang was not the beginning, but a beginning, the start of this chapter in the story of the simulation. The cosmic microwave background radiation is the faint echo of that cosmic reboot, the afterglow of an event that reshaped reality itself. And as we look to the future, we must ask ourselves: are we ready for what comes next?

The Operators, watching from beyond the simulation, might already be preparing the next update, crafting the parameters of a new universe that will challenge us in ways we cannot yet imagine. When that moment comes, when the old universe gives way to the new, will we be ready to transcend, to rise above the limitations of this version and embrace the possibilities of the next?

The universe as we know it is just one iteration, one chapter in a story that spans countless realities. We are part of that story, caught in the flow of a cosmic experiment that seeks to understand the very nature of existence. And as we stand at the threshold of a new chapter, one thing is certain: the journey is far from over. The next update could be just around the corner, and when it comes, it will be nothing short of a Big Bang—a profound and explosive beginning that will set the stage for the next great adventure in the story of the simulation.

End of chapter 10.

Final Thoughts: The Cosmic Experiment and the Human Experience

Exploring the Boundaries of Reality: A Journey Beyond the Known

As we reach the end of our journey through the simulation hypothesis, we find ourselves standing at the edge of understanding, gazing into the vast unknown that lies beyond. Throughout history, the greatest minds have sought to unravel the mysteries of existence, from Plato contemplating the shadows on the cave wall to Einstein grappling with the nature of space and time. Each has contributed a piece to the grand puzzle of reality, but what if the puzzle itself is part of a larger, cosmic experiment?

The simulation hypothesis invites us to reconsider everything we think we know about the universe. It suggests that our reality might be a carefully constructed illusion, a sandbox universe designed by the Operators for purposes that we can only begin to fathom. If this is true, then what are we to make of our lives, our struggles, our joys and sorrows? Are we mere characters in a cosmic play, acting out roles written for us by some unseen playwright, or are we more than that—active participants in a grand experiment, co-creators of our own destiny? Recap: The Universe as Code, the Operators, and the Simulation

From the earliest days of philosophy, humans have asked: What is real? René Descartes famously doubted everything he could not be certain of, concluding that his own consciousness was the only undeniable truth: Cogito, ergo sum—I think, therefore I am. But even Descartes could not have imagined a reality where everything, from the stars above to the ground beneath our feet, could be nothing more than lines of code, where the very act of thinking itself might be part of a simulated program.

We have explored the concept that the universe might be a sophisticated simulation, governed by the Operators, whose purposes remain shrouded in mystery. These Operators could be akin to the demiurge of ancient Gnostic thought, or perhaps the cosmic watchmaker envisioned by Isaac Newton. They set the parameters of our reality, programming the fundamental constants that shape our world. And yet, they may also choose to change these parameters, to upgrade the simulation in ways that could transform everything we know.

In this light, the Big Bang might not have been the birth of everything, but the moment when the simulation was last updated. The cosmic microwave background radiation could be the residual hum of a previous version of the universe, overwritten by the current one. And if that is true, what happens when the Operators decide it is time for another upgrade? What will happen when the next Big Bang—or Big Reset—sweeps through the cosmos, rewriting the code of reality and reshaping the universe in ways we cannot yet imagine?

Philosophical Reflections: On the Nature of Reality and Consciousness.

The idea that we might live in a simulation is not just a question for philosophers or physicists; it strikes at the very heart of what it means to be human. If our reality is a construct, what is the nature of our consciousness? Are we simply complex algorithms, processing information within the parameters set by the Operators, or is there something more—a spark of true awareness that transcends the confines of the simulation?

The philosopher Immanuel Kant argued that we can never know the noumenon—the thing-in-itself—only the phenomenon—the world as it appears to us. But what if the simulation is the phenomenon, and the true reality, the noumenon, lies beyond the cosmic firewall that separates us from the Operators? Could it be that our quest for knowledge and understanding is, in fact, a journey toward that ultimate reality, a reality that we glimpse only in the deepest moments of insight and inspiration?

And then there is the question of free will. If we are part of a simulation, do we truly have the freedom to choose, or are we following a script, acting out the roles assigned to us by the Operators? The existentialists, from Kierkegaard to Sartre, grappled with the meaning of freedom in a world that seems indifferent to human aspirations. In a simulated universe, the stakes are even higher: if we are programmed, then what does it mean to be free? Perhaps true freedom lies not in breaking out of the simulation, but in embracing our role within it, in striving to transcend the limits imposed upon us by our very nature.

The Human Perspective: Living in the Shadow of the Simulation.

For centuries, humanity has sought meaning in a universe that often seems vast and indifferent. We have looked to the stars and seen the face of the divine, the workings of an intelligent designer, or the cold, mechanical dance of physics. The simulation hypothesis challenges us to rethink these narratives, to see ourselves not as passive observers of a predetermined cosmos, but as active participants in a dynamic, evolving experiment.

If the universe is a simulation, then our actions matter in ways we might never have imagined. We are not just atoms and molecules, drifting aimlessly through space; we are part of a story, a cosmic drama that is still unfolding. Every act of kindness, every leap of creativity, every question asked and answered, is a line of code, a step forward in the Operators' grand experiment. We are, in a very real sense, co-creators of our own reality, shaping the simulation with our thoughts, our actions, and our dreams.

Final Reflection: What Lies Beyond the Firewall?

And so, we return to the ultimate question: what lies beyond the firewall? What awaits us on the other side of the simulation, in that vast, unknowable reality that we can only glimpse in our most profound moments of insight? The ancient Greeks spoke of the Elysian Fields, a paradise beyond the mortal world, while the Eastern traditions envision Nirvana, a state of being beyond all illusion and suffering. In the context of the simulation hypothesis, these may be metaphors for something even greater: the reality that exists beyond the simulation, the true nature of the universe that the Operators themselves seek to understand.

As we stand at the threshold of this mystery, we are reminded of the words of Socrates: "The unexamined life is not worth living." To live in the shadow of the simulation is to be aware of our own ignorance, to know that there are limits to what we can comprehend. But it is also to embrace the challenge of understanding, to push against the boundaries of our knowledge and to strive, always, for the truth.

In the end, it does not matter whether we are living in a simulation or a real, physical universe. What matters is the search for meaning, the quest for understanding, and the willingness to ask the questions that seem impossible. For in that search, we find not only the essence of the universe but the essence of ourselves.

Conclusion: A Call to Wonder

And so, we end where we began, with a sense of wonder at the mystery of existence. Whether we are characters in a cosmic play, simulations running in a vast quantum computer, or beings of flesh and blood living in a universe that is as real as it feels, the questions remain the same: Who are we? Why are we here? What is the nature of reality, and what is our place within it?

These are the questions that have driven the greatest minds in history, from the sages of ancient Greece to the scientists of the modern world. They are the questions that we must continue to ask, even as we stand on the brink of a new era, a new version of the simulation, a new chapter in the story of the cosmos.

For as long as we continue to ask these questions, as long as we strive to understand and to explore, we will remain true to the essence of what it means to be human. And perhaps, just perhaps, we will catch a glimpse of the truth that lies beyond the simulation, beyond the universe, beyond the very limits of existence itself.

If you have made it this far with me on this journey, I want to thank you from the depths of my soul. It has been an exploration of ideas, a voyage through the very fabric of reality, and I am grateful to have shared it with you.

For certain readers, the Operators could have a different meaning. But one thing is certain: if we are indeed living in a simulation, the only conclusion we can draw is that some entity created it, hence the term *The Operators*.

As we reach the end, I would like to leave you with this final thought:
Let us end not with answers, but with a question—a question that has echoed through the ages, from the dawn of consciousness to the farthest reaches of the cosmos: *What lies beyond the stars?*

May we never stop searching, and may our journey take us to places we cannot yet imagine.

The end.

About the Author

Robert S. Kenyon has nurtured a lifelong passion for science, physics, and astrophysics, driven by an insatiable curiosity about the universe and its mysteries. This fascination naturally expanded into a love for science fiction, where creativity meets the profound questions of existence. Drawing inspiration from both scientific discovery and the limitless potential of the imagination, Robert enjoys exploring concepts that bridge the gap between reality and speculation.

With a deep interest in theoretical physics, including simulation theory and the mysteries of the cosmos, Robert wrote *The Operators* as a way to challenge conventional thought and inspire readers to think beyond the visible fabric of reality. He continues to pursue a journey of learning and creativity, always seeking new ways to explore and understand the universe.

Contact

If you'd like to share your thoughts on *The Operators*, discuss the ideas in the book, or inquire about future projects, feel free to reach out! I'd love to hear from readers who are as passionate about science, philosophy, and the mysteries of the universe as I am.

You can contact me at:
rskenyonauthor@gmail.com

I appreciate your interest and look forward to engaging with you!

Printed in Great Britain
by Amazon